The floating earth : our
future in a world withou

THE
FLOODED EARTH

THE
FLOODED EARTH

OUR FUTURE IN
A WORLD WITHOUT ICE CAPS

———

PETER D. WARD

BASIC
BOOKS

A Member of the Perseus Books Group
New York

Designed by Jeff Williams

Library of Congress Cataloging-in-Publication Data
Ward, Peter D.
    The flooded earth : our future in a world without ice caps / Peter D. Ward.
        p. cm.
    Includes bibliographical references and index.
    ISBN 978-0-465-00949-7 (alk. paper)
    1. Ice caps. 2. Ice sheets. 3. Global warming. 4. Sea level. I. Title.

GB2401.7.W37 2010
363.34'92—dc22

                                                                2010000537

10 9 8 7 6 5 4 3 2 1

Perhaps the greatest threat of catastrophic climate impacts for humans is the possibility that warming may cause one or more of the ice sheets to become unstable, initiating a process of disintegration that is out of humanity's control.

—JAMES HANSEN

Everything comes to an end.

—DEATH CAB FOR CUTIE

CONTENTS

MIAMI BEACHED

Miami, 2120 CE. Carbon dioxide at 800 ppm.

Miami had become an open city. It was also an island. Although to the north it was still contiguous with the vast peninsula that had been Florida, the flooding had cut off all freeway and railroad ties, while the airport itself was now a vast lake. All this was because the level of the world's oceans had risen 10 feet. The reason for this vast geographic change—one that rendered every schoolchild's world atlas obsolete—was readily apparent. Greenland was losing its ice cover.

Now, with waters sufficiently high to render assistance enormously expensive, the embattled U.S. government could no longer bear the costs of defending the drowned metropolis against the small-time tyrants who had risen to power amid decades of economic chaos, social displacement, and political breakdown. America was in a state of triage. The nation's leaders had to decide which American coastal cities to fight for, and which to surrender to the rising waters. Miami did not make the list.

Nature was besieging the drowning city—the federal government could not afford to save a city like Miami, not with such a significant portion of the U.S. gross national product dedicated to building dikes for those urban areas of the eastern and western seaboards still somewhat less affected by the relentlessly encroaching waters. Miami had joined New Orleans and Galveston as cities given over to the rapidly rising sea and to the mobs that burgeoned in the resulting chaos. It was under no flag now. The city itself—and Miami Beach, or the thin ribbon of land that remained of it—had become a strange new entity, a variant of the old city-states of ancient Greece, but with its own peculiar problems.[1]

The sea, annually emboldened by hurricanes, had transformed the city entirely over the previous hundred years. The geography was now quite different than it had been as late as the mid-twenty-first century. To its east, Miami Beach was vastly reduced in size: Collins Avenue south to Washington Avenue was still on land, but the terrain was denuded of any but the fastest-growing vegetation and the hulks of the hotels that had once provided its economic sustenance. The vertical storm surge from hurricanes now routinely swept over the entire narrow island. The five main bridges from Miami proper, Florida highways 922 and 934, U.S. 195 and 41, and Venetian Way, had all been severed, as had the Rickenbacker Causeway, thus isolating Key Biscayne. On the other side of the onetime urban center, West Miami was the last dry land before the gigantic salt marshes that now encircled the city from southwest to northwest, effectively making Miami its own island. Miami International Airport was long gone (although its runways were still visible from the air, just under the clear water), devoured by the enlarging of Lake Joanne and Blue Lagoon along an arm from the sea parallel to the old Dolphin Expressway. Lehigh Lake had expanded northeast to meet up with the Amelia Earhart ponds to form a large brackish lake, too saline for freshwater life, not salty enough for marine life. The same was true of all the old lakes: Lake Ruth was now enlarged, with the Weston Hills the only dry land well to the northwest of the city. The numerous toxic waste sites had long ago been inundated, making the vast swamps carcinogenic until rain and freshwater hurricane flooding cleansed them of the more liquid toxins. The sediments leavened with heavy metals and arsenic were another matter.

Miami's most immediate problem was freshwater. The rising sea levels had caused the city to finally lose all semblances of municipal freshwater and sewage systems in 2106. Without freshwater from any kind of municipal system, Miamians resorted to other means. Personal swimming pools became personal freshwater reservoirs, with each residence using a combination of solar power and rainwater collectors to store water in its pool and then pump it into the house. Septic systems rarely worked well in the wet ground, so each house now had an outdoor privy, and warm afternoons filled with the stench of thousands of gallons of untreated human waste.

While the defense against the rising sea throughout the late 2000s had been heroic in its own way, the combination of loss of all personal home insurance, coupled with the crippling costs of keeping a water system running against the unceasing lateral movement of salt water into the normal aquifers and reservoirs, bankrupted the city. The federal government, with so many problems elsewhere, gave the city up as well. One relatively minor commercial decision destroyed the region's entire economic equation: the cancellation of home insurance throughout South Florida in 2073, following the enormous mortality and billions of dollars of damage from Hurricane George. Homeowners who had liquid capital (gold was the preferred coin of the realm, at $4,500 an ounce) fled the state for higher ground. Colorado was booming, for instance. Those people whose entire capital was invested in their now valueless homes (which is to say most people) either stayed and prayed or fled onto the great American road, looking for work amid the crippling national depression brought on by the battle against the rising seas. The fight for New York alone had necessitated cuts to national defense to the point that the United States had completely withdrawn from its foreign bases, defaulted on its Social Security obligations, and abandoned its short-lived national health care system. Coldly logical decision-making had led to the governmental desertion of Miami. It would not be the last such abandonment.

Here and there, the rich mansions of Key Biscayne and other upscale neighborhoods remained palaces of luxury for those whose fortunes remained. As always, enough money could still make a difference, even in a drowning city. But with no middle class to sustain a police force or any sort of municipal presence, the wealthy who remained did so with lethal defense. Money still flowed through Miami, enough to keep it from totally dying economically. But much of it came from the leading cash crop of South America, illegal drugs. Coca, that valuable crop, now covered huge regions of South America, and Miami was one of its chief ports of entry. The Miami police force had ultimately succumbed to the same realities that had brought down South American and Mexican big-city police forces in the early twenty-first century—there was just too much money, and too many well-armed soldiers employed by the cartels.

If anything, the Keys had fared even worse. All the islands were reduced in size, and a large section of the long highway connecting the mainland to Key West had collapsed, with no federal money to repair it. Key West had reverted to its old ways, once again becoming a haunt for sailors and smugglers.

There was no agriculture in South Florida at all. The rising salt levels in the soil saw to that. But perhaps the greatest change to the entire region lay to the west of Miami. Visible from the diminishing, still functioning satellites (Cape Canaveral had been leveled by the great hurricane of 2045) was the great brown-to-black smudge that had been the Everglades. Once a seemingly endless green landscape, the Everglades was now dead. It had been the first victim of the rising seawater, with all its huge expanse of freshwater plants dying quickly, while the mangroves that replaced them were still too young to make a green dent in the mass of dead vegetation. So profound was this change that the level of atmospheric oxygen over Florida had undergone a still minor but nevertheless measurable dip, joining other huge regions on Earth, such as the lower reaches of the Amazon, Nile, Mississippi, Mekong, and Ganges rivers—all these rich deltas losing all their green plants, which had been a significant portion of the world's oxygen-producing vegetation.

The sea had "only" risen 10 feet in South Florida by 2120. But the rise was accelerating. The open city of Miami was destined to become no city at all.

A FLOODING PLANET

This book is about the impending and ever-accelerating rise of the oceans due to global warming. Neither the rise nor its cause can be doubted any longer. The only question is how high the seas will rise, and how fast. But regardless of whether we reach the first 3 feet of sea level rise in 2100, or even 2200 (and not 2075 or earlier), the effects will be the same. Except with every passing month as this book is written the estimate for 2100 keeps rising. By that time anything less than 5 to 6 feet will be welcome, for the alternatives—up to 8 feet—cannot be borne by society. They are predictable because the planet's waters have risen before, even during the lifetime of our species. But in our era, it looks as though the extent and speed of rise will surpass anything that at least post-agricultural humanity has gone through. As a paleontologist who has had to professionally study the effects of rising and falling sea level from far more ancient times than the time of humanity, I know that we are not merely speculating through cloudy crystal balls; we can see from the past what we face in a future we have created. The geological record holds a rich history for scrutiny. This book is based on the fact that the earth has flooded before.

Much of what we know about the new increase in sea level comes from what we have discovered about the very old rises and falls of the sea. Recent studies of the deep past have told us much about how the earth and its life have responded to changes in sea level. Such changes in the sea level, and thus to land geography as well, have sometimes been a boon to life, but sometimes they have led to quite the opposite effect, the most extreme being mass extinction. History, then, opens one door to the explorations we will make in this book.

There are many axioms about the importance and the place of using history to contemplate current and future actions. These maxims contradict each other. We are told that those who ignore the past are destined to relive it. We also learn that the past is an unknown country, and thus a place of no relevance at all to the present. But for more than two centuries the key principle of geology has held that the changes occurring in the past are composed of processes that continue today. This is known as the Principle of Uniformitarianism, first codified by Charles Lyell, and then championed by his fellow early nineteenth-century savant, James Hutton. The principle indicates symmetry, telling us that if we can understand geological events of the past only in terms of modern-day physical processes, then the reverse must hold as well—that in terms of processes occurring too slowly to be observed in any human lifetime, the past must inform us. So it is with any change in sea level—and in the world's current predicament, the rise in sea level.

The geological record, written in rocks of many kinds and ages, tells us that the level of the sea can change in only two ways.[2] One happens as a result of the swelling or shrinking of the vast mountain chains found amid the deep ocean basins, such as the Mid-Atlantic Ridge, the long line of submarine volcanoes where hot magma erupts onto the sea floor daily and is then carried through continental drift either east or west. For reasons unknown, the heat that accompanies this ocean-long line of volcanoes waxes and wanes across the millennia. When it increases, it causes the rocks of the ridge itself to swell, decreasing the volume of the basins that hold the world's oceans. The amount of water stays the same—just the worldwide vessel holding those oceans changes in size. It is like putting a large brick in a bathtub; the level of the water rises. Conversely, with a loss of heat, the mid-ocean ridge is reduced in size, and the water level declines globally. The rate of change of sea level from this process is very slow indeed—with changes of several feet or more taking millions of years—but nonetheless significant over time.

The second method of change in sea level happens faster, coming from the accumulation or melting of continental ice sheets.[3] When snow falling in cold climes accumulates faster than it melts during the warm seasons, an ice sheet will form. All that water ultimately comes from the sea, so a growing ice sheet causes the level of the sea to drop. The opposite process also occurs: a melting ice sheet causes sea level to rise.

Long before the modern era, a time when humans began to play their own role in climate change, the seas rose and fell thanks to their own entirely natural inclinations. We can see examples of these processes in many places on the planet. Consider two American examples. The first is North Dakota, a place as far from the ocean as any place in North America—as remote, in fact, as almost any place on Earth. But a glance at the geological record of the North Dakota badlands shows that this part of the planet was once anything but landlocked.

Near the North Dakotan border with Montana, the present-day landscape is made up of the eroded beds of ancient sea bottoms—which are overlain and underlain by a different kind of sedimentary bedding, one that was deposited in river valleys. Both seabed and riverbed have their own stark beauty. They are made up of mesas and cuestas that, at sunrise and sunset, are flocked in sandstone and shale from brown to dun and seem to glow in the late-afternoon sun. The fossil and sedimentary record tells us that the lighter-colored beds were deposited in richly vegetated river valleys near the end of the Cretaceous Period, over a time interval from about 68 million to 65 million years ago, and these rocks contain the most famous of all fossils, *Tyrannosaurus rex*, as well as its more numerous prey, the herbivorous dinosaurs of the time. Yet both stratigraphically above and below the riverbeds, you can see a change in the appearance and type of the fossil content as the strata entombing them change as well. Far darker in color, these sedimentary beds are no longer the remains of river deposits, but instead are clearly the remains of a seabed, a seabed where land had been. The fossils of these darker beds have a spectacular beauty. The most common are the beautifully iridescent fossil shells of ammonites, now extinct cephalopods whose nearest living relative is the chambered nautilus of the tropical Pacific. Like a modern-day nautilus shell, these fossil ammonites sparkle in the sun when they are exhumed. But they are not alone. Among the mollusk shells are the more rare bones of giant sea lizards, Mosasaurs, as well as the biggest crocodile of all—a behemoth named *Deinosuchus,* which must have competed with the mosasaurs for food, and perhaps for breeding sites as well. Both of these seagoing reptiles had to painfully wade ashore on legs turned to flippers to lay their precious eggs, always alert for the marauding tyrannosaurs of fearsome fame.

These seabeds underlying the last dinosaur-bearing beds of North America thus tell us something profound. This far inland was once shallow ocean. It was part of the great Western Interior Seaway, a thick branch of the world ocean that existed throughout the Cretaceous—which is to say almost as long as the dinosaurs have been dead. These old strata tell us that the Western Interior Seaway was not a static entity: it rose and fell with the level of the global ocean. Thus, while the dinosaur beds of the great American West were deposited on land, they in fact sit atop the remains of the vast inland sea, one that originated more than 100 million years ago—and was caused by a rise in sea level. In this case it was not the melting of ice that raised the water, but that slight shrinkage of the volume of the ocean basins described above. At its high-water mark in the Cretaceous Period between 125 million and 65 million years ago, the North American inland sea was hundreds of feet deep and hundreds of miles across. It separated eastern North America from the western parts, creating what were really two continents. The water rose and fell, but did so very, very slowly, its oscillating history taking hundreds of thousands of years to unfold, with the final draining that resulted in our modern-day geography occurring tens of millions of years ago.[4]

But there is another place that tells us not only that the change in sea level can be vast, but also that such change can happen very, very quickly compared to those stony, primeval North Dakotan beds. That place is in a small quarry tucked into one of southern Florida's famous line of islands: the Florida Keys.

If you go down to Key Largo and don a mask and fins in John Pennekamp State Park—the first underwater park in America—and look past the 9-foot-tall, 4,000-pound bronze statue of Jesus (a second casting of the Italian bronze *Il Cristo Degli Abissi*, donated by Italian scuba entrepreneur Egidi Cressi and placed in its current location in 1965), you can behold the richness of variety of the only true coral reef in the continental United States. Its immense stone ramparts compose an invertebrate tenement occupied by untold tiny, anemone-like coral polyps, their tentacles withdrawn during the dangerous daylight, awaiting the cloak of darkness before they creep outward in search of floating meat. These enormous *Montastrea* coral are hundreds of years old and are surrounded by many other smaller corals: the blocky *Siderastrea*, the hornlike *Acropora* of two types, the more stag-like *A. cervicornis*, and a few moose-horn-like *A. palmata*, whose fronds make spiky thickets around the flank of the *Montastrea*. The corals, the *Montastrea* most of all, have lived in this region for millions of years. But now is the time to see them, because this reef is dying fast. The welcoming, warm ocean of apparent peace and biological plenty covers trouble brewing, because most of the *Acropora* corals in Pennekamp are quite dead, the ocean has been slowly warmed over the previous decades, and this Florida reef tract was one of the first to experience what would become known as coral bleaching.

The Florida sea was home to another American coral reef whose condition is a veritable graduate course in the effects of sea level increase. A short distance from Key Largo is the smaller Lone Pine Key. At one side is a flattened area that looks the entire world like an abandoned quarry (which it was). You are greeted by a mass of white rocks blazing in the sun. Up close, you'll see that the large white walls are made up of gigantic coral skeletons intermingled with a profusion of smaller coral, shell, and unidentifiable limestone crags of all sizes and orientations—a reef, a fossil reef. The largest of the fossils are enormous heads that, if you look closely, show the distinctive pattern of *Montastrea*. One of the most prominent heads is the shape of a mushroom and measures nearly 10 feet in length. This is clearly the cemetery of a reef that is identical in composition to one that exists offshore of this very key.

Yet corals of this size—surrounded and lithified with all of the framework builders and binders that make any reef, new or old, a three-dimensional, wave-resistant structure built by life and cemented by tropical ocean chemistry—develop only underwater. The components of this reef show it to be similar in structure to the offshore reefs now found at least 10, and sometimes as much as 30, feet beneath the sea. But this old reef now sits at least 10 feet above sea level. Was it lifted here by storm? By earthquake? By the simple lifting of the Florida Keys due to some unknown, deep-seated heating of the underlying earth?

The answer is simple, yet unnerving. The giant corals on Lone Pine Key have not moved a bit since the days when they were growing in water 10 to 30 feet at this very place—about 125,000 years ago. The land had not gone up. The sea had gone down. Somehow the level of the oceans—not just here in Florida, but all over the world—had risen 30 feet, stayed that way for more than a few centuries, and then receded, exposing and killing these once-lush reefs.

The world climate of our era is the aftermath of a geologically recent episode of continental glaciation informally called the ice ages, but technically termed the Pleistocene epoch. But that event, which ended about 10,000 years ago, with wholesale melting of glaciers, was just one of many that made up the ice ages. An even more significant change in climate (and in the amount of the world's ice) occurred about 125,000 years ago, when a rapid melting of continental ice sheets and glaciers caused the sea level rise that led to the formation of reefs on now solid land. These sheets must have disappeared quickly indeed, causing the world's oceans to rise and encroach on the land, carrying sea creatures with them.[5] It was not the first time such a process happened on this planet, and certainly it would not be the last, but it is the last time that a change of this magnitude happened so quickly. Now all evidence suggests that a similar rise is beginning. When the world warms, ice melts. When ice melts, the sea rises. It is this second kind of increase in sea level that promises to flood the world that our grandchildren and great-grandchildren will inherit. What might make it unique is the rate at which it is happening. There is every reason to believe that we are on the cusp of the most rapid rate of sea level rise in Earth history—that ice is and will continue to melt faster than at any previous time.

TELLING THE STORY OF THE FLOODED EARTH

In this book, I explain what the consequences might be in the next centuries. The first chapter cuts to the chase, exploring how sea level change is occurring in the present day. From there, the chapter goes into the record of ancient sea level change and the ways in which these records can be understood.

The second chapter discusses carbon dioxide—what it is, what it does, and how it contributes to the so-called greenhouse effect and global temperatures. It is the amount of carbon dioxide that is most pertinent to global climate, and every single human on the planet produces the stuff, one way or another. Hence, the number of humans on the planet is a large part of the story to come, and thus the second chapter hands off to the third, which deals with human population change and its effect on energy demand.

In Chapter 4 I look at how even a modest increase in sea level will dramatically affect world agricultural yields. In Chapter 5 I examine the confluence of global temperature increase and the makeup of continental ice sheets on Greenland and Antarctica, which hold the greatest volume of potentially meltable frozen water on the planet. I explore how and when they formed in the first place, and how fast and at what rate they will melt. When this ice melts, not just fields will be flooded, but cities as well, and that is the subject of Chapter 6.

Chapter 7 again looks back in deep time, by examining previous epochs when there were no ice sheets. Those times inevitably led to mass extinctions, and I examine whether conditions of the past leading to those mass extinctions are in any way similar to what the earth is experiencing now. This chapter shows what the very worst effect of global warming could be in the near future.

In Chapter 8 I change gears away from the science of sea level change and global warming to illustrate the potential "fixes" being proposed to remedy or manage the rise in sea level, through both local and global engineering. And I explore ways that we might yet stop the rising of the seas to levels that cause the calamitous loss of agriculture, and the equally calamitous flooding of coastal cities, and, if we cannot do that, how we might at least buy civilization more time by slowing the rate of rise. There is hope, if we act now. But the train is leaving the station. Perhaps forever.

In a way, this book is meant to be a sort of geological version of Dickens's *A Christmas Carol*. Call it "A Christmas Carol on Ice." We do have ghosts of floods past, present, and future to reckon with. The ghost of the future offers us a vision of the Chrysler Building emerging from a troubled green-and-white-capped sea, or the new island of Miami. The choice is entirely ours whether we accept that vision or create a more positive one.

CHAPTER 1

THE RISING SEA

James Ross Island, Antarctica, March 2009.
Mean carbon dioxide at 385 ppm.

There is no mistaking the sound of melting ice in Antarctica. From the pattering of individual drips to the strange groaning of the small floating slag ice—the near-final fragments of the calving glaciers—the background noise is akin to living near a freeway. It was this noise, as much as any, that brought me up from the depths of sleep, but shallow depths as always; slumber in Antarctica is a eutrophic lake of troubled dreams. After thirty-five days of hard work and close living, I was bothered by the dirtiness and personal smell and the aches from long hours excavating fossils from frozen rocks. But most of all I was disturbed about the weather, even in sleep, as I more than half-listened for the telltale first vibrations of the tent poles signaling the start of a violent and abnormal wind, a progression inevitably ending in the raucous, ear-splitting clanging of hard canvas on poles, and metal poles on each other. My partners and I had quickly learned to associate the rising wind with discomfort, fear, stress, and intimations of death foreseen. I had expected many things about Antarctica, including even these intense and disturbing storms, if not their frequency and ferocity. But I had not expected Antarctica to be melting with such gusto.[1]

Today was no day to linger in the warm sleeping bag; it was our last day, when the large ship that had landed us and our several tons of food and water on this lifeless island a month earlier would retrieve us. Already we had tempted fate enough by staying this late in the "season," the short interval when scientists could scurry across Antarctica and actually do science. In our case the science was a geological project that would increase the accuracy of dating the ancient sedimentary rocks that made up James Ross Island, a large bit of land near the end of the arc-shaped Antarctic Peninsula. We had come to collect marine fossils from near, during, and after the age of dinosaurs, with its fiery conclusion hypothesized to have attended the end of a particular asteroid's 4.5 billion years of wandering through the solar system.

But we had another goal as well. Near the end of the age of dinosaurs, the seas seemed to oscillate more than usual. The longer term rises and falls of the Cretaceous Age, such as those so starkly evident in eastern Montana and the Dakotas, were long accepted to have been caused by slow changes in the oceans' volume as the immense sea bottoms swelled or shrank in tune with the vicissitudes of deep-earth heating. Yet near the end of the Cretaceous period, the rise and fall of the sea, as evidenced in sedimentary beds, seemed to defy the rate of tectonically produced sea level change—the changes happened so fast and were of such magnitude that melting ice appeared to be the culprit. For the first time, researchers were asking whether there could have been ice at the North or South poles some 66 million to 70 million years ago.[2] That possibility had seemed laughable in recent decades for a simple reason: the entire Mesozoic Era, including the freak collision of a large asteroid in Mexico's Yucatan region that brought the period to a fiery end, was a time of elevated carbon dioxide. Atmospheric levels of this potent "greenhouse gas" were three to perhaps five times higher than the present-day level of nearly 390 parts per million (ppm). No one could conceive of ice sheets of any extent in a world with so much carbon dioxide. But could it have happened?

FINDING THE ANSWER to this question is of enormous importance. Perhaps—just perhaps—there could be much higher carbon dioxide than there is today without the resultant melting of the ice sheets and ice caps. Maybe the current rise in carbon dioxide would excuse us from such wholesale melting—and spare us the disastrous rise in sea level that would ensue. That was the good news our team was hoping for. But there were bad portents possible in the rocks we had traveled so far to see—one of which also dealt with sea level.

That an asteroid collision brought about the extinction of the dinosaurs had been unquestioned for the previous two decades. The theory had to be right: even Hollywood had taken notice by making two blockbuster movies about an asteroid striking the earth—both of them, shall we say, smash hits. In fact, the Alvarez asteroid-impact hypothesis—that the mass extinction of the dinosaurs was directly attributable to the environmental aftereffects of a large-body impact on the earth—became the dominant proposition about *all* mass extinctions; the celebrated Cretaceous-Tertiary (KT) event that did in the dinos was considered just the latest in 500 million years of animal Armageddon. But lately doubts are emerging. Coming back into consideration is a new version of a popular old explanation of animal extinction—that volcanoes were involved.[3] As a boy growing up in the late 1950s and early 1960s, I relished the wonderful movies with their clay dinosaurs jerking along in stop motion, and near the end of the movie the behemoths menacing humans would be enveloped by lava belching from a volcano. By 2008, my work and that of dozens of other geologists suggested that this might have been the only extinction even partly caused by asteroid impact.[4] All the others had another source—the explosion of volcanoes that caused a short-term version of an even direr phenomenon I was now witnessing in Antarctica. What helped to kill the earth's primordial creatures now threatens to drown us.

Long before humans were even a gleam in nature's eye, the convergence of geological forces repeatedly caused the planet to heat up. Such events, however rare, hugely altered life and its evolution. The warming had resulted from enormous volumes of carbon dioxide that emanated from the flood basalts, creating atmospheric greenhouse conditions that quickly heated the planet to a point that the poles were nearly as warm as the equator, leading the normal winds and ocean currents to diminish and in some cases totally stop. A stilled ocean, eventually even on its surface regions, loses oxygen. The apparent result was a series of nasty events, such as oceanwide "dead zones" not unlike the anoxic areas found today in the Gulf of Mexico, off Namibia in Africa, and in many lakes and estuaries where conditions of eutrophication—where a body of water first warms and then loses its oxygen as its enclosed life dies and then rots— have eliminated all the life-giving oxygen in the water.[5] Warm water holds far less oxygen than cooler water. Worse yet, warm stagnant water breeds particularly nasty microbes that metabolize using sulfur rather than oxygen. The waste by-product of that reaction was not the oxygen inadvertently produced by plant photosynthesis, but the poisonous substance hydrogen sulfide, the nasty gas that makes rotten eggs and human flatulence so unwelcome. Hydrogen sulfide is just now emerging as a new kind of dangerous pollutant. Drillers looking for freshwater are finding pockets of the stuff with sometimes fatal results, while on many beaches around the world, rotting seaweed produces local concentrations of the gas that are killing wildlife and even, in 2009, a horse on a French beach.[6] These events—whose initial actions produce toxic levels of hydrogen sulfide, and whose final actions create what is known as "greenhouse extinctions"—coincided with high sea levels.[7] The end of sea level rise seemed to coincide repeatedly with major mass extinctions, with hydrogen sulfide as one of the chief killers.

A CRY FROM ONE of my early-rising companions spurred me out of my sleeping bag. I braced myself against the usual shock of cold as I changed out of the two pairs of long underwear and thermal tops, the two pairs of thermal socks, and the thin sleeping gloves that kept me in some vicinity of warm at night. I could never get used to this cold. It had damn near murdered us on several occasions now, when high winds, horizontal snow, and wind chill temperatures made it imperative to get through the knots closing our tents in fifteen seconds or less.

Yet while it was cold enough to make us all miserable, in the summer of 2009 Antarctica was not cold enough. I was seeing with my own eyes the evidence of a study authored by Eric Steig, the man in the tent next to mine. Steig had found that Antarctica, like every other continent, was warming, releasing its ice into the world's waters.[8] His finding inspired the wrath of Senator James Inhofe of Oklahoma (the leading Congressional skeptic on global warming) to denounce Steig as a scientific fraud. He accused Steig and his coauthors of somehow making up the data, while all around our campsite, even in the placid, ice-studded bay in front of us, melting rapidly proceeded.

The second night I was there, I'd endured the evidence of global warming's impact on Antarctica. A great storm had descended. The next dreadful morning sent me groping for the safety of the food tent in complete whiteout, when I foolishly tried to go for food rather than wait out the storm and try to hold my tent poles against the wind. In the cold, trying to decide if I could find either my own tent again or the food tent, quickly getting really cold and terrified, I bumped into Steig, who turned me around to face the direction he knew the food tent to be.

We made it there, to find the rest of our seven-person team already sipping coffee and tea, each with a story of the extraordinary storm. Our camp manager, an employee of the National Science Foundation who had organized and orchestrated numerous field camps such as ours in Antarctica, was downplaying the storm, as any old hand invariably does. Not above 35 mph winds, she said with a straight face. Steig and I exchanged a look. Both of us knew storms and wind; we both live in Seattle, which has had its share of near-hurricane windstorms, and this wind was a lot closer to 100 knots than it was to 30. Just an average Antarctic storm, she said. What the wind tried to do to our tents had nothing to do with normal weather or simple equipment failure. It was all about the reality of the changing intensity of Antarctic storms. Later we would hear that the storm had totally wrecked the camp of another paleontological party on a nearby island, sending them half frozen to a providential cave for two days until a ship could rescue them. Their tents had ripped to shreds in the wind. Argentinean scientists camped across the island from us later told us it was the most severe storm any of them had known in more than thirty years of annually coming to this part of the peninsula, with its sustained winds of 70 knots, and gusts far above that. Steig told me it was the most violent storm of his Antarctic experience— multiple expeditions that already totaled years, not months, of continuously living here.

The cause of those storms, anomalous warmth, was evident in more than the newly violent weather. Our camp was a swamp, even well into the Antarctic autumn. The permafrost on which our camp was placed was melting—perhaps for the first time in 10,000 millennia. Between and beneath each of our tents was an enlarging pond; we waded from place to place in our camp. Getting around required rubber boots.

The ship was there, all right, silhouetted and dwarfed against a towering wall of ice taller than a ten-story building; across the bay, where the ship now gently rocked amid floating ice, was a giant lobe of ice, a fragment of the huge ice sheet that covered part of James Ross Island. It was spectacular—as were the waterfalls it had gestated, all of that long-frozen water entering the sea. I wondered if there was an instrument sensitive enough to measure the increase in sea level produced by our own island's melting ice sheet. Our small island's contribution to the level of the sea, even if all of its ice were to melt, would probably raise the world's oceans less than a few millimeters. But multiply that by all the enormous ice sheets on this planet, each melting, and you get a flood. A flooded earth.

THE WATERS UPON THE EARTH

The subject of this initial chapter is nothing less than the overall subject of the book: that the level of the global ocean can change, has changed in the past, and is changing again. The rates of change have varied from very slow to very quick. ("Glacial" is an inapt term, since change in glaciers is fast compared to tectonic rates.) The alarming aspect of this otherwise normal effect on the earth is that today's predictions of sea level increase all appear to be too conservative. What makes things alarming is that this time, the rates of change seem anomalously fast compared to those of the past.

The first victim of sea level change will be geography—at least as we know it. Growing up, all of us learned early the basic geography of the world. We may not have learned the names of all the countries on that map on the schoolroom wall, but the world's continental configurations were imprinted on our young brains. We knew the outlines of the great landmasses, the larger islands, the overall Rorschach-like inkblots of green, yellow, and red land floating in blue sea. Because we learned them so young, the shapes of the continents feel not just familiar but permanent. In fact, before 1960, all school kids were taught that the shapes of continents and oceans had been the same through all of time. It was only relatively recently that the continents were discovered to be inveterate travelers upon a spherical globe, all the while being fixed to the earth's mobile, if stony, upper crust. But we will have to change our maps long before the continental plates creep significantly across the planet. The shapes of coastlines are set to radically change beginning early in the next century, as water moves quickly to resculpt and cover the low-lying land areas we are familiar with.

What causes sea level change? As noted in the Introduction, the oceans are affected by the machinery of Earth's solid surface mass, including both tectonic processes (mountain-building that causes some land to rise, some to fall), and the phenomenon of isostatic rebound, where land springs upward if a heavy load of long-term glacial ice is removed. Another factor affecting local sea level occurs when a coastline gets a load of sediment rapidly dumped on it from rivers, from the flow of debris from coastal mountains, or when strong long-shore currents subside from the weight of the sediment. Sometimes the settling is of greater magnitude than the offsetting rise of the bottom caused by tectonic uplift. This event is especially true for deltas, where rivers meet the sea and form broad aprons of low-lying sediment over areas dictated by the strength of the river and the load of sediment it carries. For example, coastal subsidence in any river delta region is estimated to be 10 millimeters per year, which will help minimize the effects of sea level rise.

Oceans rise more in some places than in others.[9] Although we intuitively consider an upsurge in sea level to be a global phenomenon, because the world's oceans are all connected and it seems as though every part of the "bathtub" would fill up at the same rate, on a spinning planet with great differences in heat from equator to pole such complex workings as oceans vary considerably. Relative sea level changes might be either higher or lower than the global average. A number of geological and oceanographic factors contribute to this geographical variability. Despite the great distance, the relevance to the United States is unequivocal. The rate of rise will be higher on the East Coast than the West Coast, and it is the East Coast that has the largest American population at risk.

Although measurable, tectonic events rarely precipitate large-amplitude changes in sea level. But changes far more rapid than any caused by tectonic forces are evident, even since the end of the ice ages some 10,000 years ago. That there are so many flood myths is no coincidence, and what would a brief rise in sea level be interpreted as *other* than a traumatic flooding, whose major consequences would rattle down through the ages as sad tales of tragedy?

So many myths—but until recently, so little proof. The problem with proving this contention was the great difficulty in first finding geological evidence of major sea level change since agricultural times, then pinning down the exact time it happened. With the advent of new dating techniques, however, as well as a far more sophisticated means to interpret the ancient sedimentary record, the mystery of at least some of these flood legends was solved.

Narratives of water covering the land are indeed found in many cultural histories. The story of Noah's flood is one of the most widely quoted legends of scripture, telling of a rise in the level of water over the entire earth, resulting from monstrous, continuous rains lasting almost six weeks. Science tells us it is impossible to rain everywhere on Earth at the same time, and even if it were possible, the high mountains of the many continents would never be covered completely. To cover the earth to 29,029 feet—the highest point on Earth—would take as much or more water as the oceans contain already. Yet the Old Testament tale of Noah is just one allusion among civilizations after 10,000 BCE attesting to an event that could have been brought about only by a rise in sea level.[10] Both ancient Greek and Middle Eastern literature recorded accounts of a mighty flood, while the Sumerians told of a similar event in their region.

What, if anything, really happened? Perhaps the flood myths originated in a truly spectacular event that occurred in the Black Sea some 7,600 years ago. In 1998, geologists Bill Ryan and Walter Pitman found evidence of a short-term rise in the Black Sea that would have flooded enormous areas on its entire perimeter.[11] As much as 60,000 square miles might have been flooded in as little as a few months. This event, so short, so obviously devastating, could indeed have started the myths of a global flood that comes down to us from multiple ancient sources in that region.

The obvious question is what could have caused such a short-term rise. All the water spilling over the normal edges of that landlocked sea had to come from somewhere. Was it new water suddenly added by the many rivers that flow into the Black Sea? That explanation simply begs the question of why the rivers would so suddenly increase in volume. Was it a truly spectacular deluge of rain, dwarfing anything we've seen today?

The answers to this riddle come from fossils found in the region. Before the great flood, now many thousands of years ago, the Black Sea was no sea at all but a lake, perhaps the largest lake on Earth. Immediately after this event of about 5600 BCE we see a very different kind of life entombed as fossils around the Black Sea: organisms adapted to salt water.[12] The rise of the Black Sea could only have occurred by flooding from the global ocean in some way. Somehow the waters of the nearest ocean, the Mediterranean (and by extension the Atlantic Ocean), made their way into the Black Sea.

The event had to have been spectacular beyond television disaster-movie fare, because the only entrance would have been through what is now the Bosphorus Strait but was then a valley, near what is today Istanbul. Water would have cascaded through that valley from the Mediterranean into the Black Sea in a volume perhaps two hundred times greater than the present-day flow of Niagara Falls.

Yet that one mystery solved leads us to another: why would all that salt water so suddenly make its way inland? Perhaps a titanic earthquake opened a passage from sea to lake. But perhaps something else was involved—a rapid increase in the level of the global ocean, which would have had to rise for some reason. But before we look at explanations for this phenomenon, we should examine what the two geologists found within ancient rocks. Such evidence is the key to understanding the variations in sea level past, present, and future.

The increase in the level of the Black Sea, it turns out, did far more than flood nearby land areas: it would have substantially changed the nature of sedimentation in the region. For starters, there would have been a change in the shorelines. Shore deposits show a characteristic pattern of sedimentation and sedimentary structures such as wave or ripple marks, even dunes. In addition, human settlements would have been flooded, and even the kind of sediment falling onto the deepest parts of the Black Sea would have changed.

Other evidence showed that the flooding was substantial—to the point that the level of the Black Sea may have risen as much as 558 feet (although others think it may not have been more than 200 feet).[13] Either way, the Black Sea is in a confined basin, surrounded by higher ground, and the rising sea would have covered all the agriculturally rich lowlands very quickly. However much the world's ocean rose overall, it produced enough new water to severely flood this region.

There is no doubt the world's oceans rose about this time, for good reason. About two hundred years before the Black Sea flood, global temperatures shot upward. This short-term rise in temperature, about 7,600 years ago, is called the Holocene thermal maximum—the warmest period of the past 10,000 years (before our own, that is). The temporary heating of the atmosphere caused ice caps on Greenland, and perhaps in Antarctica, to melt at least partially; it also spurred the melting of many mountain glaciers in lower latitudes. All that water had to go somewhere—and it ended up in the global ocean, causing it to rise, and sending water spilling into the lake that became the Black Sea.

Noah and the geological record tell us that sea level can change, and relatively quickly. The fastest way is through the melting of ice. And melting ice is precisely what is going on right now.

As recently as 15,000 years ago, the contours of the continents looked very different because of these rapid changes—caused by the melting or formation of ice caps.[14] In the twenty-first century, we have just come out of an ice age where sea level was 240 feet below what it is now, and that translates into far greater continental land areas lying exposed. New Guinea was hooked to Australia by dry land; North America, especially along its tectonically quiescent East Coast, was far broader than today; the Mediterranean was more a lake than a sea. If we considered each second as a century, we would see the onlap and offlap of the seas, the draining of gaping "bathtubs" like Hudson Bay, and the refilling and gobbling up of coastlines as the sea returned back onto the land. In fact, all occurred at a rate that would have been noticeable in a single human life span. That is the kind of change we face now: a rate too slow to be perceived by politicians elected to two- or to six-year terms, but a rate too fast to leave civilization as we know it unscarred.

THE RATE OF THE RISE

The greatest single scientific question—and for our society, a question of life or death—is how far and how fast the seas will rise. There is no end of prognostications. It has been estimated that more than a million Web sites discuss the issue, many of them proposing their own guesses. While plenty of uncertainty remains about how extreme the rise in sea level will be, the exact rate will make a vital difference in who lives and who dies. A rise in the earth's temperature of just 7 to 9 degrees Fahrenheit will not by itself kill anyone if it happens slowly. Nor would a 30-foot rise in sea level necessarily be lethal, because the amount of time it would take to occur would allow even the slowest creatures to crawl their way to higher ground—except when storm surge batters down dikes, potentially killing thousands at a time. But the consequences of 30 feet—or even 10 feet—are staggering, simply for the amount of land either covered or affected by storm surge and a process known as salt intrusion, where salt-enriched water seeps sideways into the soil, ruining it for human crops or even native vegetation. The loss of arable land alone from these two processes guarantees massive famine on an unprecedented scale. Animals would easily walk away. Plants would try to disperse. But human-held property does not get up and move toward higher ground. Someone loses, someone pays.

So what are the estimates of sea level rise, and how do we assess them? The best data at hand come from careful measurements of current sea level rise. That rise seems minuscule, on the order of several millimeters per year. It also seems to have been constant throughout the twentieth century. If sea level rise were to stay at present levels of increase per year, there would not be any risk of catastrophic land loss by incursion of the sea for tens of thousands of years, if ever. But no one believes that rate of rise will remain at present levels.

It doesn't take much of a change in climate conditions to edge us from manageability into catastrophe. As we saw in the Introduction, based on evidence emerging in Florida, only slightly more than 125,000 years ago the global level of the sea was about 13 feet to almost 20 feet higher than it is today. The beautiful fossil corals now found well above sea level in the Florida Keys are testament enough: such corals are found far off-shore, in depths of 10 to 30 feet. There were no Florida Keys then, no Key West, no Everglades or Miami Beach. All were underwater.

One vital, frightening factor unites today's world with the radically transformed one that existed 125,000 years ago: carbon dioxide levels in the atmosphere were lower than those of today, but not by much. Florida and places like it were submerged by seas when global temperatures were only about 4 degrees Fahrenheit higher than they are now, and most of that rise came from the melting of the Greenland continental ice sheet—not the ice already floating in water, but the ice on land. This difference is crucial. We hear about the loss of Arctic sea ice. When that melts, sea level does not change. But the ice on land that melts goes right into the sea—and raises its level. Since that time, there has been no geographic change of Greenland, such as a move northward by rapid continental drift. Greenland's continental ice volume melted before under conditions that we are fast approaching (or have surpassed) in terms of global CO_2 and global temperature. That means we can expect a similar rise of perhaps 30 feet even if we succeed in keeping carbon dioxide levels no more than about 500 ppm (390 ppm as I write this in 2009).

Today we are bombarded by so many numbers, and estimates of sea level change are part of this resulting explosion of information—much of it contradictory. Yet the various recent estimates of sea level change mostly seem to discount the observations made from deep time—of changes so great that they would utterly topple the map industry with the rapidly changing geography of sea level rise. Among the many data being published are many "official" estimates of sea level that purport decreasing sea levels, and foes of global climate change have trumpeted this decline. But a study in late 2007, based on rates of increase during this most recent rise, projects that by 2100 sea level will be almost 6 feet higher than today—and still rising. Following this, the Copenhagen meeting of late 2009 again raised the estimate—to a *minimum* of 5 feet! This is a far higher number than that estimated by the United Nations Intergovernmental Panel on Climate Change (IPCC), which lists maximum, minimum, and estimated ranges of sea level change.

So what else happened 125,000 years ago? Four papers published simultaneously in 2006 evoke a world with radically different coastlines than we have now. The first of these papers used computer simulations to derive a picture of what the Greenland ice cover would have looked like and how big it was. The researchers found that most of the ice fields in Arctic Canada and Iceland had disappeared and that the Greenland ice sheet had been reduced to "a steep ice dome" in the central and northern parts of the country. These results squared with other evidence about the ice core and with other paleoclimate data, all of which agreed that much of southern Greenland had been entirely deglaciated—it was bare rock where today a giant lid of ice exists. All that water went into the oceans, where it created higher sea levels. And yet with much of the southern ice fields of Greenland gone, sea level was higher than would have been expected even with such extensive melting.

A second paper tried to explain why sea level became so high 125,000 years ago, suggesting that it was not only Greenland that was melting, but parts of the Antarctic ice sheets as well. This paper reckoned the rate of sea level rise at that time was at least 10 mm per year—which conservatively means a full 3 feet of increase per century. But such figures are not the highest rates of sea level rise in the past—far from it, in fact.

Understanding that climate change is under way certainly predates the start of the twenty-first century. By the 1980s and 1990s, scientists worldwide had become alarmed enough that they scrambled to start regular meetings and, more important, begin the difficult job of convincing their national governments that this newest assault on civilization warranted action. One of the results was the formation of the IPCC, composed of an international group of scientists of broad backgrounds. Their meetings and consultations have resulted in the publication of ever-thicker tomes attempting not only to summarize changes that have taken place in recent years but also to attempt to forecast and predict future rates and trends of climate change. Two of the most important issues are future temperature change and future sea level rise.

Although the IPCC is avowedly nonpolitical, internal politics surely played a large part in the ultimate results of its four reports so far. Thus there is much conjecture by the scientific (and political) communities on just how confident we can be in its various estimates. It is probably safe to say that while the estimates are accompanied by weaseling about high and low possibilities away from the mean estimate itself, surely a dose of conservatism enters into the projections. No one wants to be associated with some out-of-bounds and wild overestimate.

Because sea level change is related and correlated to temperature change, the two estimates from the fourth IPCC report (IPCC 4) have been rigorously examined in print and on the Internet since their publication in 2007.[15] Let us look at what the IPCC foresaw, and how close to reality those estimates might be.

The first of the IPCC 4 estimates of sea level rise led to widespread confusion in the press. The media have extracted a positive message from the estimate, though it is certainly realized that there is no good news in any aspect of sea level rise; thus the lower the estimate, the better the news.[16] Journalists had the impression that the estimate touted by the fourth IPCC report was actually less dire than the group's previous accounting. The critical figure relates to how many inches (or millimeters) sea level will rise by 2100. Some articles came out reporting that the new estimate was of a 23-inch rise, instead of the 35 inches estimated in the third IPCC.[17] This would indeed be good news—if sea levels rose only 2 feet over the next century, surely humanity would be able to compensate and engineer itself out of the kind of disasters a more rapid rise would cause. But were these press reports accurate? The very complexity of the IPCC estimates may in fact have led reporters to misinterpret the findings.

Estimates like those presented by the fourth IPCC report are based on mathematical models, and all involved various scenarios of climate change.[18] It is important to see exactly what these models represent. Only then will we understand what these estimates really mean and how valid they may be. At the heart of all this is that climate change is highly politicized, and that the politics involved in the IPCC estimates may have trumped the science.

Let us look at what the models show, with the caveat that model "results" are at best scientifically informed estimates, not real data. The first IPCC 4 model explored what is called "thermal expansion." Warming water expands; if the entire ocean warms, enough of its water molecules interact to cause a rise as the entire volume of seawater slightly expands. But this is not a straightforward scenario. The sea is warmed at its surface, and if warming were limited to this region, there would be far less thermal expansion. But the numerous ocean currents, including the so-called thermohaline systems—the vastly important and gigantic currents that carry surface water down into the depths at various places around the globe and oxygenate the ocean bottom—also serve to import warmer water from the surface to the depths. The rate of this warming has to be taken into account as we model how much the ocean temperature changes. The models have to estimate where, and how fast, the warmer surface waters penetrate the depths.

The second part of this modeling exercise related to the rivers of new fresh water entering the sea from melting glaciers and ice caps—excluding the great ice sheets of Greenland and Antarctica, which are so important that they merit their own modeling. Scientists seek to estimate how much of the ice resting on land surface is disappearing—or will disappear—into the oceans as the world warms. They compute this number from formulae linking mean global temperature to loss of ice—which itself is an estimate. Hence we see one of the major problems of this strategy: using estimates to make further estimates.

The third IPCC modeling section related to Greenland and Antarctica and estimated the rate at which their vast ice sheets would melt as temperatures increased. This calculation involved a number called surface mass balance, or the amount of snowfall (which causes ice accumulation) minus ablation (melting) of the ice. This modeling shows whether the glaciers are getting larger or smaller, not so much from their length, but from their thickness.

Finally, there is a fourth way the ice sheets can contribute to sea level rise, one that is a bit tricky to comprehend, even counterintuitive, yet it must be evaluated in model form as well. Ice has the unusual property of being able to essentially "flow" while still remaining frozen. Thus if the rate of such flow increases, more and more of it hits the ocean, where it melts, and sea level rises. It is this kind of flow that particularly worries climate scientists, because it can happen very fast—and can change quickly. We can envision the "melting" of glaciers on land in two ways. Yes, they are melting in place—ice turns to water, and that water makes its way to the base of the glacier and then flows out beneath the rest of the glacier into the sea. But the entire "river" of ice is also heading toward lower ground, and eventually the ocean if it is near enough to the sea—which most of Greenland's glaciers are. Huge blocks of ice cleave off the front of the glacier and fall into the sea, where they melt, and mix with the other water that came from beneath the glacier.

To come up with an overall estimate of sea level change, the IPCC ran each of the four models on computers, then examined the results. The IPCC 4 came up with the following estimates of how much the seas would rise overall by the years 2090 to 2099, as compared with sea levels from 1980 to 1999.[19] The various scenario numbers certainly add to the confusion (a specialty of the IPCC reports, in my view—why not just scenarios 1 through 6?). Each of the scenarios modeled slightly different conditions of overall global warming, from coolest to warmest.

B1 scenario: 0.18–0.38 m rise in sea level (or a maximum of less than two feet)
A1T scenario: 0.20–0.45 m
B2 scenario: 0.20–0.43 m
A1B scenario: 0.21–0.48 m
A2 scenario: 0.23–0.51 m
A1FI scenario: 0.26–0.59 m–or a maximum of just over 2 feet

In the most extreme case, the last one in the table above, sea level would rise less than 3 feet by the end of the twenty-first century. Hence the science press published the huge sigh of relief after the IPCC 4. The drop from a maximum of 0.70 to 0.59 m between IPCC 3 and 4 in the estimate of the rise of sea level seemed good news. But almost immediately after the publication of the IPCC 4, new studies of *past* sea level rises—examining not a theoretical future change but increases that really did happen—put the IPCC estimates into question. Moreover, scientists actually working on the ice sheets themselves were very worried about IPCC projection of the rate of flow of the ice sheets, a process poorly known indeed. In the face of rapidly rising temperatures, they asked, could ice flow increase to the point that the ice sheet actually disintegrates into a catastrophic, off-the-charts melting? Such a cataclysm cannot be ruled out, and in fact, in late 2009 such a catastrophe was discovered to have happened in Antarctica about a million years ago—the first proof that ice sheet disintegration is not just theoretical (but more about this event in a later chapter).

Beyond these projections and the challenges to them, there is a good current estimate about how fast the oceans are rising within their basins. It is a seemingly very small number: less than 2 mm per year. Yet there is something disturbing about these numbers, because IPCC's models woefully underestimate current sea level change. A good way to test the models is to use them to predict something with a known answer, such as the rate of sea level rise this decade. Given this exercise, the models "flunk." In other words, when the numbers directly observable from modern-day melting and temperature are plugged into the models, the figures that emerge from the many equations seem to be way too low. The models thus appear to be wrong, in that they project melting at rates that are only half of what can be measured today.

Based on all of this number crunching (and political posturing), it is instructive to hear from climatologist Gavin Schmidt, one of the authors of the IPCC, about what the real rate of sea level rise could be:

The main conclusion of this analysis is that sea level uncertainty is not smaller . . . , and that quoting the 18–59 cm [a meter is a bit more than 3 feet] range of sea level rise, as many media articles have done, is not telling the full story. Fifty-nine centimeters (59 cm) is unfortunately not the "worst case." It does not include the full ice sheet uncertainty, which could add 20 cm or even more. It does not cover the full "likely" temperature range that global warming could produce by 2100 (up to 6.4 °C)—correcting for that could again roughly add 15–20 cm. It does not account for the fact that past sea level the models underestimate rise for reasons that are unclear. Considering these issues, a sea level rise exceeding one-meter can in my view by no means be ruled out. In a completely different analysis, based only on a simple correlation of observed sea level rise and temperature, I came to a similar conclusion. As stated in that paper, my point here is not that I predict that sea level rise will be higher than IPCC suggests, or that the IPCC estimates for sea level are wrong in any way. My point is that in terms of a risk assessment, the uncertainty range that one needs to consider is in my view substantially larger than 18–59 cm.
—GAVIN SCHMIDT, RealClimate.org

As a final point, let's remember that all these projections of sea level rise focus only on an increase over the next century. Estimates stop a hundred years out. But rising sea level will not stop. Over the coming centuries, we can expect meters of sea level rise at a minimum—and much more if something unexpected happens, such as rapid disintegration of one or more ice sheets.

HOW HIGH COULD THE SEAS CLIMB?

So with this entire circus of estimates to contend with, how fast could the oceans rise? It certainly seems that the changes produced by melting ice can produce rates of change higher than those powered by tectonic changes, such as greater heat flow into oceanic basins causing thermal expansion of the ocean basins, and thus lowering their overall volume. Thus we need only look to the time of the great ice ages to find maximum rates. While there are numerous times of melting to choose from, it seems as though one of the fastest sea level rises on record occurred during the most recent ice melt, 14,000 to 16,000 years ago.

To study this rate of change, climate scientists needed to have a means of measurement. One such means was to locate positions of mangrove and tidal flats that had left behind evidence of their existence in the sediment. But because sea level was rising, the shoreline was also migrating. Taking a single drilled core of nearshore sediment, for example, one finds that the contact between sediments deposited on land area that was subsequently flooded would be meaningless—that land surface had to be tracked horizontally as it migrated landward, in retreat from the advancing sea. The deposits left behind by mangroves, which, it turns out, are readily identifiable in cores, had to be tracked over large geographic distances. Scientists undertook this investigation on the Sunda Shelf area of Southeast Asia.

The rate of rise turned out to be *nearly 50 feet over three hundred years*—or more than 15 feet per century. A large portion of an ice shelf or sheet in Antarctica appeared to have rapidly disintegrated, causing wholesale melting—enough to cause a catastrophic rise in sea level. Thus, as shown throughout this chapter, we see that sea level rise becomes the single greatest natural danger threatening civilization as we know it.

Fifteen feet in a hundred years would surely have been alarming to the many humans living by the edge of the sea, who were not only watching the waters rise but also witnessing radical changes wrought on the river mouths, estuaries, and salt marshes, and the region's near-shore agriculture. The sea would have swallowed entire villages over time. Now there are cities by the sea, with far fewer villages. Picking up stakes and moving to higher land is a lot easier to do with a village than a concrete and steel city with all its infrastructure.

THE RISING THREAT OF STORM SURGE

Although the ascending sea is perhaps best imagined as a slowly rising if relentless tide, in fact there is a second manifestation of rising sea levels that happens fast. Among the more rapid changes, the most ominous is produced by great storms hitting coastlines where sea level has risen to new and higher levels than human civilization has engineered for. This process is called storm surge, and it threatens a great number of humans.

Storm surge happens when large storms essentially push great volumes of water up against the shore, where the water can then break on the shorefront as abnormally high waves (which can move farther inland than most other waves). The most recent estimate is that more than 200 million people worldwide live along coastal flood plains that can be brutalized by storm surge, representing about 4 percent of the world's population.[20] (These areas lie below what is termed the once-in-1,000-year elevation that surge flood might attain—during the worst storm of a millennium.) Of this number, the climatologist Neville Nicholls—who has done more than anyone in showing the effects of sea level rise on the world's future— estimated that, on average, *10 million people per year* experienced flooding due to storm surge in 1990.[21] But any rise in sea level changes that 1990 equation, as does any increase in human population in the low-lying areas next to the sea. Both the sea and the flood-plain populace are rising, making storm surge ever more dangerous to ever more people. But it is not just people who are threatened; storm surge can take out local agriculture and infrastructure, such as roads and railroads.

Some important geographical variations in storm surge are evident when we view it from a global perspective.[22] As might be expected, at greatest risk are the small, low-lying islands, such as those in the Caribbean Sea, the Indian Ocean, and the Pacific Ocean. Among continental regions, by far the most vulnerable to increased surge flooding are Europe's southern Mediterranean, West Africa, East Africa, South Asia, and Southeast Asia. It goes without saying that this same group of coastlines will similarly be disproportionately affected not just by storm surges but also by future sea level rise.

Increasingly, as we perform the all-important calculus of cost-benefit analyses around whether we fight sea level rise or withdraw our battlements and human settlements from the coastlines, we will have to factor in storm surge. Whether on the municipal, state, or national level, our planners tend to make their assessments on topographic contours, establishing their strategies based on land they believe will be above sea level in 2100, the farthest date that any planning group seems able to imagine. But storm surge can upset all their best-laid plans. In the years ahead we will face two kinds of flooding. The first is the kind of catastrophic submergence that will occur when once and for all the sea rises onto a given height above (old) sea level. The second is the more transient kind—the flooding caused by storm surges. Such large storms will become ever more dangerous with even small escalations in sea level. For low-lying, densely populated regions, such as the Bay of Bengal and Holland, storm surges pose the greatest threat to life.

Storm surge is often portrayed as the waves crashing into a coastline amid rain and wind. Wind-whipped waves are indeed the major destructive elements caused by seawater (or freshwater in large lakes). Yet although there are indeed large waves in a storm, storm surge itself is a temporary rise in sea level as more and more water is pushed landward. At the intersection of land and sea, the water has no place to go but up. Adding insult to injury, the large waves roll in on the back of a raised sea level.

Whether combating storm surges or hedging against the more subtle and long-term effects of rising seas, civil planning is always based on probabilities and is an exercise in what technocrats call "risk management," which, in truth, is condoned gambling. The gambling comes from the hope that any structure built to tolerate some kind of periodic catastrophes, such as river flooding or hurricane damage, can withstand various disasters for its lifetime. All engineering structures have a functional lifetime, and engineers work from the belief that the sites of their structures experience no more than the strongest possible storm only once every ten years or once every fifty. But increasingly, Mother Nature fools the engineers by throwing a once-a-century storm at structures not built to withstand it. Such was the planning in New Orleans before Katrina—the levees were designed to withstand the once-a-decade category 3 hurricane, not a once-a-century category 4 or category 5 storm like Katrina.[23] They gambled and lost.

While this kind of speculative engineering has, in fact, paid off in the long run, it was introduced in situations where the danger to be encountered occurred at some unknown but presumably constant rate. Buildings, levees, roads, bridges, and other structures have been designed to deal with other constancies. Flood engineers do not worry that the Mississippi River will keep rising, year by year, eventually overtopping its banks even during non-flood times. But that is exactly the situation facing the structures that have been built to withstand the ravages of great storms hitting coastlines. As they confront the reality of storm surge, climatologists and oceanographers have been able to arrive at tables estimating the maximum surge heights—the rise in effective local sea level caused by surge. Surge height is important because it creates a short-term but real rise in sea level of varying heights. Everyone concentrates on the permanent rise. But the occasional rise in major storms is equally important, or more so. Say you are a kid with a big fence keeping you from the forbidden but wonderful goodies of a next-door orchard. You know that sooner or later you will grow big enough to hop that fence. But then someone gives you a stepladder . . . and the goodies that the storm surge—the stepladder for rising sea level—gets to are our cities and farms.

Surge height is affected by the geography of a coastline, such as a funneling bay, or a large intersecting river that could add water, or even by the topography of the sea bottom just offshore. Worst-case scenarios for surge heights occur when a storm encounters a topography that induces strong waves during maximum high tide. Many bays experience a double whammy, such as the Bay of Fundy in Canada, where waves from 33- to 50-foot tides can produce a devastating effect.[24]

With the recognized acceleration of annual sea level rise, atmospheric models and oceanographers have begun to examine how much higher storm surge might get as the world warms and ice melts. While any rise of carbon dioxide (the subject of the next chapter) does nothing to the level of the sea, the warming that carbon dioxide produces will certainly melt portions of the great ice sheets of Greenland and Antarctica as well as the many mountain glaciers around the world, whose melt water will eventually reach the sea by rivers. A 1997 German study by H. Von Storch and H. Reichardt predicted the rise of storm surge levels with higher CO_2 levels along the country's long coastline. Other researchers followed suit for many other coastlines around the world, and the results are sobering.

The Von Storch and Reichardt study tried to estimate the storm surge the German coastline would encounter when atmospheric CO_2 had doubled. With CO_2 at 385 ppm now, and climbing at 2 ppm per year, the level of 770 ppm would occur in only 190 years, or about 2200—assuming the rate of yearly CO_2 increase did not accelerate. In this calculation, we have to consider the increase in sea level from other factors, such as the melting of ice in Greenland and Antarctica, and the melting of mountain glaciers in the face of such high concentrations of greenhouse gases. Because most studies conclude that sea level will rise a minimum of about 3 feet by 2100, a minimum estimate for 2200 would be 11 feet—but more likely will be 5 to 30 feet *at a minimum!* The storm surge calculations for both the German coastline as well as selected other European coastal cities varied from slightly less than 3 feet in additional rise (for Bergen in Norway, and Galway in Ireland) to a whopping 18 feet (for the German town of Esberg; England's Liverpool fell in between at 5 feet). Esberg could conceivably see a sea level rise of 25 to 35 feet, even using conservative figures.

Another worrying aspect is that CO_2 will rise faster than IPCC 4 conceded in 2007. An extreme view is held by David Battisti of the University of Washington, who projects a CO_2 level of 800 to 1,000 not in 2200, but by 2100. Although this early arrival of such high CO_2 will reduce by a century the amount of time that melting significantly raises sea level, Battisti also points out that current Global Circulation Models (GCM) underestimate the effects of such levels of CO_2 on most Earth systems, including melting, and thus sea level. So while the IPCC takes comfort in projecting "only" about 3 feet of rise at most in sea level in the year 2100, it did not add in the storm surge number of an additional 3 to 10 feet globally. Finally, the warming atmosphere will surely increase the number and ferocity of storms hitting coastlines. Opposition to the hypothesis that global warming increased hurricane numbers and strength in the last parts of the twentieth century and first decade of the twenty-first century seems to be melting away.

Without a doubt, worldwide, there will be far more incidents of coastal flooding than currently predicted. Because of these effective heights with the addition of storm surge, it is probable that great swaths of low coastal countries will be inundated, and even if the sea subsides back into its basin, the damage to agriculture, infrastructure, and other developed human property will be enormous. As we will see in later chapters, Holland and Bangladesh would fare the worse, but other places, notably Venice in Italy, will probably be made uninhabitable and will have to be abandoned. We could witness inundated cities and coastal communities that produce nothing but consume much, and that would put brakes on their country's GNP. "Recession" is too fine a word for what would ensue from storm surge alone; major economic depression would be probable.

As I write this, a new cascade of doubt flows through the press. Predictably, conservatives such as George Will dismiss those of us who, like the great Peter Finch in the movie *Network*, are screaming out of windows through books such as this. We are labeled Cassandras.[25] I am not sure what a Cassandra is. But I know what I indeed am: scared.

SEA LEVEL RISE AND "SCIENTIFIC RETICENCE"

Science is built on facts. But to no less an extent it is built on reputations. Scientists routinely put greater or lesser weight on a new scientific pronouncement based on the perceived accuracy and believability of the scientist behind it. It is quite true that reputations can be inalterably affected not only by presenting poor data and poorly drawn conclusions, but also from crying wolf or even exaggerating effects of good data. "Scientific reticence" is a well-known phenomenon defined by philosophers of science as "resistance by scientists to scientific discovery." In my own field, paleontology, this phenomenon was in full swing in the decade following the announcement by the Alvarez group from Berkeley that the dinosaur-killing KT mass extinction of 65 million years ago was caused by the environmental effects of a large asteroid hitting the earth, rather than by intrinsic causes, such as climate change or disease, which had been the favored reasons for a century up to the 1980 discovery of iridium and glassy spherules at KT boundary sites that helped verify the impact extinction hypothesis.[26]

The phenomenon of scientific reticence is in full swing around the debate over sea level change, present and future. Although vertebrate paleontologists viscerally opposed the asteroid-impact hypothesis out of fear they would back the wrong horse, in the case of climate change, the forces resisting estimates of sea level increases are more complex and subtly different. Now the fear is not so much of tarnishing one's reputation by adhering to one particular hypothesis, but of crying wolf in response to new discoveries—that is, publishing information that ultimately turns out to be wrong to the detriment of a scientist's career. Even the investigators who make the new discoveries are reluctant to sound the alarms their findings demand. NASA climatologist James Hansen, who has pointed out wholesale examples of scientific reticence concerning many aspects of climate change, sees the current climate change situation as causing many scientists to be more worried about crying wolf than "fiddling while Rome burns."

While not overtly saying so, Hansen is really accusing his colleagues of scientific cowardice in the face of what he sees as overwhelming crisis. He has addressed this issue in two ways—starting out by being gentle:

I believe there is a pressure on scientists to be conservative. Papers are accepted for publication more readily if they do not push too far and are larded with caveats. Caveats are essential to science, being born in skepticism, which is essential to the process of investigation and verification. But there is a question of degree. A tendency for "gradualism" as new evidence comes to light may be ill suited for communication, when an issue with a short time fuse is concerned. However, these matters are subjective.

But in the same paper, he took the gloves off:

I suspect the existence of what I call the "John Mercer effect." Mercer (1978) suggested that global warming from burning of fossil fuels could lead to disastrous disintegration of the West Antarctic ice sheet, with a sea level rise of several meters worldwide. This was during the era when global warming was beginning to get attention from the United States Department of Energy and other science agencies. I noticed that scientists who disputed Mercer, suggesting that his paper was alarmist, were treated as being more authoritative. It was not obvious who was right on the science, but it seemed to me, and I believe to most scientists, that the scientists preaching caution and downplaying the dangers of climate have fared better in receipt of research funding. Drawing attention to the dangers of global warming may or may not have helped increase funding for relevant scientific areas, but it surely did not help individuals like Mercer who stuck their heads out. I could vouch for that from my own experience. After I published a paper (Hansen et al. 1981) that described likely climate effects of fossil fuel use, the Department of Energy reversed a decision to fund our research.[27]

Hansen makes no bones about his own estimate: that sea level will rise more than 3 feet by the end of this century. He goes on to say that this estimate itself may be conservative, and as noted above, so it seems—now we must hope for less than 5 feet of rise.

Ultimately, as we move into the second decade of the twenty-first century, what is the best estimate science can produce of how high and how fast the world's waters will go? It is certain that sea level will continue to rise—for rise it has, globally, for some time now. The amount it rose over the twentieth century has been small. But will that smallish rise stay so—or, in the face of the accelerating increase in atmospheric greenhouse gases, will it begin some inexorable increase, both in rate and in amount of land it swallows? Those are the critical questions that in no small way control the fate of coastal humanity.

Science never stands still, however, especially when confronting a crucial issue like sea level change. But because we are dealing with effects in the future, we must always work with estimates, not data. All our guesses come from models—but that is the best we have. Most recently, in the late summer of 2008, a new set of estimates emerged, based on a new kind of model. Climatologist Stefan Rahmstorf used a statistical protocol to ask how high sea level will be by 2100—compared to known values of 1990.[28] His new method assumed that the rate of sea level rise was related to, and mathematically proportionate to, the rate of warming above the temperatures of the pre-industrial age. The method can be tested using real data—by matching the real rate of rise during the twentieth century with real data on CO_2 as well as sea level rise. With these in hand, Rahmstorf then modeled the future. There are always uncertainties in modeling, and thus good modelers do not try to foist a single number or result on their audience, assuming they have one. A better methodology is to come up with minimum and maximum possible results. The caveat here is that the amount of temperature rise must be estimated for the future in order for the model to work.

The results are not encouraging for anyone wanting the world's oceans to stay right where they are. Employing the known data from the twentieth century, Rahmstorf established that the sea rose 3.1 mm for every degree of temperature rise in excess of prehistorical values. By using the estimated temperature rise assumed for the next century, he came up with an estimate of sea level rise from 1990 to 2100 to be a minimum of 2 feet to a maximum of almost 5 feet. As we will see in the next chapter, 5 feet above where we are now will cause major changes to our world.

CHAPTER 2

RISING CARBON DIOXIDE

Athabasca region, Canada, 2030 CE. Carbon dioxide at 420 ppm.

I t seemed as if no place was immune from the seas of the past. The oceans had their influence even here, east of the Rockies, where the vast Canadian prairie appeared even more endless than its southern extension, the Great Plains, in the United States. Every river exposed sedimentary rocks that told the same tale—that long ago, this place was under water. Even with the most extreme possibility of worldwide sea level rise—the 240 feet promised if all of Antarctica and Greenland were to melt—this prairie would remain dry. Yet it was here, in this part of Canada, perhaps more than any other single place, that could incite that kind of catastrophic melting. Athabasca was a hotly contested place, one where sandstone richly endowed with tar-like carbon had become, by 2030, the greatest single producer of carbon dioxide.[1]

From Calgary, Alberta, the road to Athabasca traversed a prairie of annual grasses swaying in the constant wind, a landscape formerly cold most of the year because of the high latitude. But now its escalating warmth made it increasingly valuable for wheat and other grains. Even farther to the north, the landscape was rapidly changing, moving from a place with an occasional withered pine to a terrain of numerous small trees growing lustily in the ever-longer summers with their endless twilights. The land gave way to creeks, then ponds and even lakes, then immense rivers that drained north, not west, making their stately way into the Arctic Sea. Mosquitoes darkened the sky from May through October. This region, part of the vast Northern Boreal Forest, a huge expanse formerly dominated by jack pine and black spruce, now experienced the northward march of new kinds of trees, moving with the changing climate. Broadleaf trees could now hold their own.

Warmer or not, it was a difficult land to traverse because of the muskeg—swampy, wet land that would break the leg of livestock or of the unwary human hunter. The wildness was palpable: the rivers and lakes held schools of arctic grayling, with their sail-like dorsal fins, while river otters trolled for shellfish, and black-throated green warblers flitted among the trees. If you were wary and careful, you could even sight an endangered species, the whooping crane, its immense wings causing their own muted but audile roar on takeoff, while woodland caribou watched warily from dark copses of boreal forest—evidence that the protections put in place during the twentieth century may have actually done some good. The land had looked this way since the last ice cover of the ice age had finally melted away to the north, leaving behind a rolling landscape littered with gravel and glacial features that eventually evolved into a soil cover, then adapted for cold, and now was unfolding through the novel warmth. It was a beautiful land, increasingly rich, because the new warmth increased the rate of weathering—the decomposition of earthly matter from interaction with the atmosphere—which in turn produced more and better soil. It was a land inhabited for at least 10,000 years by various tribes of indigenous peoples.

But as you approached the Athabasca region, all began to change. Soon the surroundings no longer resembled green Earth at all but became a reasonable facsimile of the lunar landscape. Plants were gone; jumbled piles of sand, muck, and rock were piled in vast disorder; and there were even craters—the vast, man-made settling ponds. Even the bedrock beneath the land had changed: the vast, fine-grained, and light-colored sandstone cropping out along the riverbanks and eroded knolls had been replaced by a darker rock, one that was greasy to the touch and smelled of oil if broken open. This was tar sand, a thick sandstone infused with rich hydrocarbons forming a mass called bitumen. This rock type was found at or near the surface of the region over an area exceeding 54,000 square miles, in three separate deposits. Taken together, these Athabasca tar sands made up one of the largest known oil deposits in the world, second only to the fabulously rich sandstones of Saudi Arabia. In 2030 as in the immediate decades before, oil fever ruled this part of Canada. As in any gold rush, the dream of wealth trumped law and culture. Athabasca's First Peoples, having lived here 10,000 years, had witnessed the effects of a gold rush on land they formerly owned. Now they were in the middle of a guerrilla war.

In 2009 the First Nations communities downstream from the existing tar sands petroleum extraction plants were labeled a national security threat to energy production in Canada, allowing the government of then prime minister Stephen Harper to initiate plans originally created for pacifying villages in Afghanistan. In particular, when the Mikisew Cree and the Athabascan Chipewyan protested the poisoning of their waterways and the health and environmental hazards facing their peoples, they were "monitored" by a government determined to wring its oil from the land at any cost.

To the north of Athabasca, a new toxic waste "lake," devoid of life except for microbes and undeserving of its own name (although the locals had dubbed it Lake Harper, after the leader who had seen to it that Canada's long practice of environmental stewardship was put aside for the Athabasca treasure), continued to grow beyond its 75-square-mile size. By 2030, it contained enough toxic water to top off all of Lake Erie by a foot.

Even back in 2009, doctors attending the First Peoples noticed among their patients an alarming number of rare cancers and autoimmune diseases such as rheumatoid arthritis and lupus. The cases were too numerous to ignore even given the small population. Already in 2009, populations of moose living near the settling ponds were found to have five hundred times the "acceptable" level of arsenic in their meat. The health and environmental fallout of oil development incited an incendiary fury among the First Peoples. As early as 2020, the Fort Chipewyan and Fort McKay communities had begun actively sabotaging the oil fields and had started kidnapping oil field workers—following the lead of equally angry locals near Nigerian oil fields. To the northeast of the Athabasca region, the

Lubicon First Nations held up the construction of the Gateway Pipeline, which was to cross from the tar sands to Kitimat in British Columbia for shipment to China. By 2030, the Canadian army now had a real role in defense—defense of the tar sands.

Along with the immediate human health hazards in Athabasca arose a more insidious threat—one that would affect people far beyond the region's borders: the creation of carbon dioxide. Converting tar to oil released carbon dioxide, and lots of it.

The development of this rich reserve had taken place for decades. To exploit it, giant bulldozers stripped away the thin soil of glacial till, and then other great machines scooped up the tar sands and dumped them into waiting trucks. Athabasca's Syncrude mine had the dubious honor of being the largest mine of any kind in the world, and as the other petroleum fields of the world were pumped dry one by one, over the first decades of the twenty-first century the Canadians increasingly exploited the tar sands.

Converting the raw rock into usable petroleum took energy. Like the old saw that it takes money to make money, here it took vast amounts of energy to end up with the liquid gold. Barrel by barrel, the tar sands became petroleum. But mining this material required not just fossil fuel and electricity: water was the other essential ingredient. Water was needed to convert tar sands to oil, but even more water was needed for the slag heaps of useless gravel generated by the conversion process to try to make the waste less toxic, for it still contained enough hydrocarbons to make people sick if it got into groundwater. It needed to be cleaned and isolated because the by-products of the mining became settling ponds that were enormous and abiotic. Nothing but strange metal- and oil-loving bacteria could live or grow in otherwise sterile bodies of water such as Lake Harper.

All of this effort resulted in large quantities of two very different products. The first was a liquid filled with carbon compounds that burns when heated: heating oil and gasoline. The second was a simple molecule: carbon dioxide, a carbon atom with two accompanying oxygen atoms bonded in place. At Athabasca, it was produced in its most common state, as a gas. As is true in one way or another of most sites our species inhabits, carbon dioxide was emitted from the trucks and machines, and from the cooking and heating of the workers' accommodations during the frigid winters and short but hot summers. But here in the tar sands, by far most of the carbon dioxide came out of gigantic chimneys rising from the vast refineries where the solid oil shale was turned into the usable liquid. Visible even from space, these chimneys had a most dubious honor, one that in fact they shared with the entire region around the mine: more greenhouse gases were produced here each day (and night, because the mining and refining never stopped) than anywhere else on the planet.

In 2030, there were innumerable signs that the world's ecosystem was subtly changing, its temperatures rising, its sea levels creeping higher. Such heavily populated spots as China, South Africa, and South Australia suffered from continued droughts. But because none of these changes had yet caused any sort of mass human mortality, little was being done to ameliorate them. The industrialized countries would meet, haggle about reducing carbon dioxide emissions, and then go home with nothing accomplished. Three decades into the new millennium's oil-based economy, there was no way to seriously reduce emissions without grave damage to the economies and agriculture that were barely keeping the 7 billion people on the planet fed and clothed.

Of all the changes, the biggest of all was invisible to everyone but the scientists and high-altitude animals and plants. In 2030, for the first time in millions of years, carbon dioxide stood at 385 ppm, near the so-called tipping point identified by many in the early years of the twenty-first century. To the many millions born in those years, 385 ppm seemed like status quo. It had been at least a decade since journalists and politicians had tried to stir in the multiplying billions of humans any sense of global apprehension about escalating carbon dioxide levels. In fact, the situation was now akin to the first six months of World War Two in Europe—the time of the "Phony War" that soon enough ended in bloody battles. By 2030, the increasing number and ferocity of hurricanes, the ever-greater flooding, and the continued disappearance and widespread, complete loss of mountain glaciers had all happened as predicted, but other events always seemed to divert humanity's attention elsewhere. And one factor had not come into any significant play at all: since the start of the century, the level of the sea had risen only about 100 to 150 millimeters—a paltry 4 to 7 inches. Skeptics had a field day on that aspect of global warming. Little did they know that major change was coming to the seas as well.

Even as they did nothing, all in this world agreed about one thing. It was definitely getting warmer, and that warming was mainly because of a simple molecule, one absolutely necessary for the continuation of life on Earth: carbon dioxide. This rise in carbon dioxide had lit the slow fuse that was starting a melting that would end in Greenland and Antarctica when every bit of ice was gone.

From molecule to melting: the cause and effect between carbon dioxide and wholesale sea rise was evident now, its implications irrefutable.

THE LIFE AND TIMES OF CARBON DIOXIDE

In September 2009, the British government released a report on the newest models forecasting how much and how fast global temperatures could rise in this century.[2] The results upended previous estimates suggesting that this century's rise in temperatures would not be catastrophic and that none but today's newborns would be around to feel the coming heat—even as teenagers or young adults, the Class of 2028, or even 2038, would graduate from high school in a world little if at all different from present-day Earth. The new report, produced by the United Kingdom's Met Hadley Centre for Climate Prediction and Research, declared that average global temperatures could rise by 7.2 degrees Fahrenheit by 2060. These startling new predictions came in the aftermath of yet another major letdown for progress on the issue of climate change at the autumn meeting of the G-20, the score of nations with the largest economies. The Met Hadley Centre's study, conducted on behalf of the UK Department of Energy and Climate Change, anticipates a 10-degree or greater rise in temperatures in some world regions, which would cause droughts in certain areas and flooding in others, severe rises in sea level, and the collapse of entire ecosystems. The Arctic could be up to 15.2 degrees Fahrenheit warmer, while the temperature rise in Antarctica would be nearly as large. The consequences of these changes are stark—but why are they happening at all, and so quickly?

According to most experts, the reason for this coming (and current) warmth is increased carbon dioxide. Nearly all scientists agree that human activity is causing CO_2 levels to rise, though a few hold out that today's levels are the result of natural cycles unrelated to what we have done to the planet, and that we need do nothing toward curbing emissions. Those who would deny global warming contend, without convincing evidence, that the rise in global temperature is in fact the result of long-term climate cycles, or a consequence of changing solar activity operating on a time line invisible to our science because we have not been looking long enough. But their minority view is totally debunked by the deep time record of past CO_2 and past climate changes. However, it does not seem like a minority view if you tune in to politically conservative radio shows and telecasts. In this chapter, I'll demonstrate that rising temperatures—and the ensuing rise in sea level—are indeed a consequence of rising carbon dioxide. To do that, I must take a detour away from the sea itself and focus instead on the planet's atmosphere—and not just today's atmosphere but the ones that characterized Earth eons ago. Just as the geological record gives us invaluable information about past sea level change (that seas do rise and fall, and that we can estimate the speed at which the process happens), so too will we better understand greenhouse gases if we examine their record in deep time. That way, we will learn much about why CO_2 is warming our planet and raising our oceans.

THE PAST TELLS US MUCH about our present—but it can tell us much about our future as well. The near-future scenario of a CO_2-spewing Canada warming the entire world is hardly a fantasy. Athabasca is doing more than any other single site to heat our skies and thus ensure that our oceans encroach on our lands. But it is just one human enterprise among many that is introducing unprecedented levels of carbon dioxide into the atmosphere and disrupting the delicate balance of earth, sea, and sky.

Is carbon dioxide a villain? This relatively rare and otherwise innocuous gas molecule now seems to rank right up there with humankind's worst bogeymen, real or imagined. Yet its rise to notoriety is recent: even a decade ago, CO_2 was just another minor gas, not considered a primary factor in the dynamics of climate, and certainly not judged as a pernicious atmospheric force. How things have changed. Now it appears to be the central player in what is truly a long-running drama, that of climate change.

All the carbon found on, in, and above the earth today was formed in some ancient star, in the dense fusion furnaces contained in all stars' interiors.[3] As the solar system emerged from the early solar nebula, the inner planets formed with heavier metals on average, while beyond the orbit of Mars the planets received very little of the heaviest elements and became instead great gas giants. That there is so much carbon and water on Earth is testament to the great volumes of outer solar system material brought sunward by comets and asteroids, with most of it crashing onto the various moons and planets from Mercury through Mars. As the earth finally formed and cooled, it found itself covered with seas and a thick atmosphere of nitrogen, carbon dioxide, and water vapor. But the carbon in that CO_2 has not remained in gaseous form all this time. Instead, its bonds with the two oxygen molecules would be readily broken in many different chemical reactions, which frees the molecule to pass between various "reservoirs": the oceans, the solid earth beneath us, and living matter. It can manifest itself as a solid, liquid, or gas. Thus while carbon dioxide has been present since the earth's inception, it is extremely doubtful that many of the CO_2 molecules now floating in our atmosphere are original to our planet.

The first real scientific look at the elusive gas phase of carbon dioxide was undertaken by a reportedly dour Scotsman named Joseph Black in 1754.[4] By putting limestone into a weak acid and carefully recovering the gas bubbles fizzing off the stone before they dissipated in the air around him, Black soon discovered that nothing would burn in what was to him a new gas. He also found it denser (that is, heavier for the same volume) than air. But Black's greatest contribution was to show that carbon dioxide was the gas emitted by animals—including humans—during respiration. Black, a practicing physician, wanted to know how human breathing and metabolism worked.

A good start, but there were other properties of the gas still to be discovered. One key finding was that carbon dioxide, if concentrated enough, is toxic, even lethal, to humans. In 2010, our atmosphere contains carbon dioxide at a rate of nearly 390 ppm. If that number were to rise to 10,000 ppm, or still only about 1 percent by volume of the atmosphere, once again we would be at a truly ancient atmospheric concentration, one that may have been present before the evolution of animal life. If CO_2 were to rise an order of magnitude higher (such as 10 percent of the atmosphere), animals would pass out and die within hours. It was exactly this kind of death that was suffered by more than 1,800 people living on the shores of Lake Nyos in Cameroon in 1986. Over the years volcanic forces beneath the lake, itself a large volcanic crater that had flooded, discharged gases, including a great deal of CO_2, that immediately dissolved into the lake water. Eventually, however, the water could absorb no more CO_2 molecules, and like any unhappy landlord, it began evicting the freeloaders. But rather than emitting the gas a small volume at a time, the lake burped an enormous bubble of CO_2. Because carbon dioxide is heavier than air, this moving cloud of nearly pure CO_2 stalked the shoreline, enveloping people and animals, snuffing out their lives in great silence.

CARBON DIOXIDE AND NATURE'S GREENHOUSE EFFECT

Carbon dioxide in the atmosphere is vital to life as we know it—even though too high a proportion of the gas could suffocate the very life it otherwise engenders. Its primary property is the ability to trap heat and infrared energy close to the planet's surface. This is the famous greenhouse effect, although as we'll see, this appellation is inaccurate to describe what CO_2 really does in the atmosphere. No atmospheric gas acts in the same way as windowpanes in a real greenhouse (or in any house, in fact). The real process is similar but different, the distinction subtle but important. Yet the first understanding of the greenhouse effect seemed to have spawned from a seemingly unrelated phenomenon: the first construction of glass greenhouses for plants.[5]

The first greenhouses (which were quite primitive by our standards) intended specifically to help grow plants were built in seventeenth-century Italy by the benefactors of another breakthrough: the first affordable glass to appear in the marketplace. Anyone walking into a greenhouse on a cold but sunny day immediately perceived the warmth within, and from this came the notion that perhaps gases alone could do the same to the entire planet. By the time of John Tyndall, in the 1860s, greenhouses ranged in size from small to large, and by large I mean on a grand scale indeed. Experimentation with the design of these first greenhouses in seventeenth-century Europe led to marked improvements as technology produced better glass and improved construction techniques. The greenhouse at the Palace of Versailles was an example of their size and elaborateness; it was more than 500 feet long, 42 feet wide, and 45 feet high. But perhaps the greatest of all greenhouses was the Crystal Palace of London.[6] Built in 1851, it was 1,851 feet long, with an interior height of 108 feet, and sported all kinds of interesting exhibits and innovations, including the first 3-D renditions of dinosaurs (posed to look like squat lizards), then but newly discovered.

With the sudden increase in the number of greenhouses for growing tropical plants in cool, damp England—from tiny domestic structures all the way up to warehouse-size transparent castles—naturalists discovered firsthand how glass could amplify the sun's output, both from the blazingly bright light beams themselves and from the air warmed by convection and turbulence, which are forces causing gentle spinning and overturn of air masses of different density within the greenhouses. With the giant greenhouses still fresh in their minds, scientists soon discovered that some of the atmosphere's minor gases played a major role in climate. Although the actual term "greenhouse effect" would not be coined until 1896, the overall acceptance that warming was occurring through the action of minor gases had been prevalent for the previous fifty years.

But a greenhouse is a misleading metaphor to describe the role of CO_2 and other gases in our atmosphere. In fact, carbon dioxide, methane, and water vapor (along with several other gases at even lesser concentrations) cause warming in ways quite unlike the warming within a glass greenhouse. Nevertheless, the thought that a gas as well as glass could warm an enclosed space (in this case the lower atmosphere of the whole planet) was born. This idea that carbon dioxide and a few other gases could, even at small concentration, trap heat within the atmosphere and work as a kind of natural greenhouse goes back to the 1820s, when the great naturalist and mathematician Joseph Fourier realized that energy in the form of visible light from the sun penetrates the atmosphere to reach the surface and heat it up, but that this same energy (now converted to heat) cannot so easily escape back into space. The heat rising from the warmed surface of the earth is absorbed by air itself as it tries to bounce back into space. This process is something like a one-way mirror for visible light—it passes through as it would through a window in one direction, but bounces back as it would in a mirror in the opposite direction. The equations and data available to nineteenth-century scientists were far too poor to allow an accurate calculation. Yet the physics was straightforward enough to show that a bare, airless rock at the earth's distance from the sun should be far colder than the earth actually is.

As great a mathematician as he was (to this day the Fourier Transform is used in statistics), Fourier could not derive the correct equations that would allow him to model this so-called greenhouse effect, and he was even more hopeless as an experimenter. Nearly forty years passed before Tyndall used experimentation to show that the greenhouse effect was produced by several gases in addition to CO_2, including methane and water vapor. But Tyndall made a crucial mistake: he thought these gases worked like a single pane of glass, and in this he was influenced by the increasing number of glass-and-iron greenhouses so popular in Europe and America in his time. The great Swedish scientist Svante Arrhenius termed the phenomenon the "greenhouse effect."[7] Furthermore, he was among the first, if not the first, to posit that the combustion of fossil fuels might increase the level of CO_2 in the atmosphere and thereby warm the earth several degrees.

The warming process occurs like this: energy in the form of visible light from the sun easily penetrates the atmosphere to reach the surface and heat it up, but this energy, now transformed to heat, is stuck there (because of one of the great laws of physics: energy can be neither created nor destroyed, but transformed). Unlike light, heat cannot so easily escape back into space. The air (nothing more than a swarm of nitrogen and oxygen molecules) absorbs the heat (a process technically called infrared radiation) rising from the earth's surface. The heated air radiates some of the energy back down to the surface, helping it to stay warm. This process is actually quite different from what happens in a greenhouse, where the glass holds within it the warmed air as a single blanket. But the atmosphere is infinitely more complex than a single, blanketing pane of glass. To understand what is happening in our atmosphere, we must complicate that analogy to imagine *multiple* layers of air, like multiple electric blankets, with air spaces in between, on a bed. But in this case the warming is not from electricity, but from light. The greenhouse effect might be a metaphor limited in its scientific authority, but it is one we can use if we recognize that the process it describes is the result of many factors at work.

In addition to keeping the planet warm enough for animals and higher plants, CO_2 is also the major source of carbon for life. But what I find surprising as a scientist is not that greenhouse gases can warm the planet by trapping heat in the atmosphere as well as provide the major scaffold molecule of life, but that the greenhouse gases do both with relatively few molecules. If we could randomly take 1,000 molecules of gas out of our atmosphere, 790 would most probably be molecules of nitrogen, and 210 would be oxygen (there would be variations in such an experiment, but if enough experiments were performed these values would predominate). But for all of these tiny molecules now in our sampler, in all probability there would be *no* greenhouse-gas molecules at all! While we measure nitrogen and oxygen as parts per hundred, we measure carbon dioxide in parts per million. Going back to our gas capture experiment, in most of our experiments to get a single molecule of carbon dioxide we would have to collect 2,500 molecules of gas. Yet that one molecule, surrounded by 2,500 nitrogen and oxygen molecules, when combined with its brethren, is enough to warm the earth more than about 50 degrees Fahrenheit on average above what it would be if there were no greenhouse gases at all. Left without CO_2, our planet's average temperature would thus be just slightly below 0 degrees Fahrenheit—more than cold enough to freeze the oceans. Without this gas, the surface of our planet would be too cold to allow most animal and plant life, and plants would not have produced the oxygen we breathe.

Yet for all its beneficence, more than a little CO_2 can be too much of a necessary thing. Far greater dangers can be associated with this molecule than those that come from several billion years of life-giving atmospheric gaseousness, or even from the stealthy peril arising from a volcanic lake brimming with CO_2. Rising levels of carbon dioxide could cause swift global warming and concomitant loss of oxygen in the oceans. I call this mechanism a "greenhouse extinction," and I will discuss it in Chapter 7.

THE HUMAN ROLE IN THE RISE OF CARBON DIOXIDE—AN INTRODUCTION

During the first half of the twentieth century, the scientific community produced ever more mathematically precise modeling of the greenhouse effect. Less studied, however, was the source of the gases themselves. Where was CO_2 coming from? Water vapor was one easily understood source—the evaporation of seas, lakes, and rivers. It took brave souls to stick gas sniffers into the maw of various volcanoes to ascertain the kinds of gases produced there. But most early researchers thought humans were unlikely to substantially affect climate through their own production of carbon dioxide. Because the oceans contain fifty times as much CO_2 as the atmosphere, and the concentrations of CO_2 in the oceans and the atmosphere seemed (at least in early measurements) fixed, it appeared that perhaps only 2 percent of the human-produced CO_2 would remain in the atmosphere.[8] Unfortunately, this happy assumption was not to last. In 1957, climatologists Roger Revelle and Hans Seuss demonstrated that the oceans could not absorb CO_2 as rapidly as humanity was releasing it, presciently noting that "human beings are now carrying out a large-scale geophysical experiment."[9] After this revelation, monitoring stations were set up to measure worldwide trends in atmospheric concentrations, and by the mid-1960s, it was clear that humanity was rapidly changing the atmosphere through its increase in resident greenhouse gases.

Climatologists now believe that a doubling of CO_2 would warm the earth 3 to 8 degrees Fahrenheit, which could leave our planet warmer than it has been during the past 2 million years or more. Moreover, other human-produced gases entering the atmosphere also combine to increase the greenhouse effect, including methane, chlorofluorocarbons, nitrous oxide, and sulfur dioxide.

But to understand—and address—what our species has done to aggravate the level of carbon dioxide in the atmosphere, we need first to comprehend how carbon dioxide has manifested itself in our planet's atmosphere over the past nearly 5 billion years.

CARBON DIOXIDE OVER TIME

Carbon dioxide has been a vital part of the earth's atmosphere since the planet's formation, some 4.6 billion years ago.[10] The level of this gas has varied through time—a variance that has had important and far-reaching effects on the biosphere and on evolution. The main effect produced by varying CO_2 levels comes from its well-known, quite natural greenhouse effects: during periods of relatively higher CO_2 levels, the earth will be warmer than during times with lower levels.

While it is easy to measure carbon dioxide in the present-day atmosphere, assessing the levels of CO_2 that existed in ancient times has proven difficult. We can directly measure fossils and extrapolate from the carbon isotopic ratios of ancient soil nodules, but most of what we accept about past CO_2 values has come from mathematical modeling. The most recent graphs show a kind of yo-yo. From 600 million to about 300 million years ago, CO_2 levels were far higher than today—perhaps as high as 1,000 to 3,000 ppm. But then in the geological period called the Carboniferous Era, CO_2 levels plunged even lower than in the recently concluded ice age. By 250 million years ago, however, they went back up to the highest levels found earlier. The final change has been a long-term slide back down to our present-day levels.

New data have allowed a more precise set of estimates of past CO_2 levels at various times in Earth's history. My own work on mass extinctions of the deep past showed an apparent relationship between CO_2 and the deaths of species. Even more important, another stark bit of evidence looms, proving that the times of high carbon dioxide were also times of elevated sea level. Times when there were no ice sheets.

If it were only a coincidence that high carbon dioxide levels accompanied the mass extinctions and high sea levels of the past, then we might be okay despite the impending spike of CO_2 to 1,000 ppm. If, however, as I argued in *Under a Green Sky*, high carbon dioxide produces global warming that can occur so fast that it kills off entire species, then we will soon witness a mass extinction. How soon such an extinction might start is the next pressing question. Would it happen over the next thousand years, or hundred years, or a few decades hence? We will return to this question in more detail in Chapter 7.

Even if we avoid a mass extinction in the near term, what will the world be like with carbon dioxide at 1,000 ppm? Ice caps and ice sheets have been part of our planetary ecosystem for the past 35 million years. New findings indicate that CO_2 levels dropped from about 540 ppm to a near-modern level of 370 ppm between 40 million and 30 million years ago, and then continued to drop below the 200s during the Pleistocene, the great ice age of the past 2.5 million years.[11] But then a change occurred. Levels started to rise, as they continue to do today. How far and how fast will they ascend? The most extreme estimate suggests that within the next century we will reach the level that existed in the Eocene Epoch of about 55 million to 34 million years ago, when carbon dioxide was about 800 to 1,000 ppm. This might be the last stop before a chain of mechanisms leads to wholesale oceanic changes that are not good for oxygen-loving life.

THE CLIMATE ENIGMA OF ANCIENT "GREENHOUSE" WORLDS

We are certainly not in the middle of another ice age, where large areas of continents even to midlatitudes (such as to the upper Midwest of North America) are covered by a mile of ice, and sea level is hundreds of feet lower than it is today. Nor are we experiencing what we might call a full greenhouse climate, where continental ice sheets, if they are present at all, cover only the highest latitudes, and where sea level is tens to hundreds of feet higher than now. But both of these rather dramatic end-of-the-curve manifestations of climate can provide interesting clues about what might be in store.

My own career has been dominated by my research into the Cretaceous Period—the time extending from about 135 million to 65 million years ago. The third of the trio of geological periods making up the middle era of animal life known as the Mesozoic Era, the Cretaceous Period began with something of a whimper. Unlike many of the geological units,

its start was defined not by a great mass extinction, but instead by a relatively minor event, one hardly recognizable on land, and marked in the sea only by rather minimal changes to the marine faunas then living: a few thousand species died out, rather than the millions of a major mass extinction. Its end, however, was anything but minor—a great asteroid struck the Yucatan Peninsula, killing off a majority of species, including the iconic dinosaurs.

The Cretaceous can be characterized as having been one long summer.[12] It had some of the highest sea levels in geological history, which meant that continental ice sheets were either minor or absent, or that the ocean basins were warped upward in spectacular fashion by greater-than-normal heat flow from beneath, a process we saw earlier that causes a reduction of volume of the ocean basins without any lessening of the volume of seawater filling those basins. All that excess water had to go somewhere, and that somewhere was the margins and even interiors of the Cretaceous continents. There is little evidence of the latter, and much evidence of the former. Earth was a warm place, too warm, perhaps, for ice of any kind. It was so warm, in fact, that it has been dubbed a "hothouse" climate—a time very particular and peculiar in the past. Perhaps only the Eocene Epoch (of the Paleogene Period, some 56 million to 50 million years ago) was equivalent. We seem to be returning to this climate, where Arctic temperatures were above freezing (perhaps there were a very few exceptions) and there was no floating sea ice at all. Crocodiles frolicked in what we now call Hudson Bay—which also suggests that there may have been no ice there. It looks as though Antarctica was also ice-free. These times are of relevance to us because of their high atmospheric carbon dioxide—levels that our world may once again experience within centuries or less at current rates of increase.

So just what was this kind of climate like? That question has been of considerable interest to climatologists, and a frequent subject of recent debate. Previously climatologists took comfort in believing that tropical regions during the Cretaceous and Eocene were not much warmer than today (suggesting that our tropics are already as warm as they can get). Data for Cretaceous- and Eocene-age carbonates—analyzed for their ancient temperatures by comparing two isotopes of oxygen, one with a molecular weight of 16, by far the most common, and the far rarer O^{18}—can show the temperature at which an ancient lime rock solidified. This method revealed that the ancient super-greenhouse worlds of the Cretaceous and Eocene had tropical sea surface temperatures no warmer than those of today—85 degrees Fahrenheit. Recent data now contradict this. Better samples that have been shown not to have been secondarily deformed, heated, or otherwise compromised by Mother Nature and her very tectonically active earth have given a more ominous view of ancient tropical heat.[13] The tropics back then did in fact heat to temperatures hotter than those of today, as evidenced by sea surface temperatures as high as 95 degrees Fahrenheit measured from sedimentary rocks of that age. That is pretty warm water—so warm that modern-day corals would have a hard time surviving for long without undergoing a process called bleaching, in which their symbiotic algae leave for cooler climes, leaving the coral animal white or transparent in appearance, and in so doing leaving it with a death sentence as well.

Where does this new information leave us? It appears that there really are two climate regimes—the world with ice, and the world without. For the past 2.5 million years our planet has not been able to decide which way to go—the climate has bounced back and forth between times of much ice on the continents, and warmer times, with less ice. But left to its own devices, the earth's now historically low levels of planetary CO_2 (rarely below 300 ppm in the past billion years or more) probably would have kept us with continental ice sheets for many tens of millions of years hence. It is entirely plausible that in the course of natural history, CO_2 would not—until about 7 billion years from now, when our sun will become a red giant star and utterly destroy the earth—reach the 1,000-ppm mark, its average for the past 500 million years. But humans have changed that history.

Our future promises a change in carbon dioxide levels initiated by humans but not directed by them. The combined climate and sea level change are like a big boulder at the top of a steep mountain. Anthropogenic greenhouse gases—that is, gases generated by humans—have already sent the rock rolling. Down it goes, and we can do nothing to stop it. Temperature will increase by 3 to 4 degrees, and sea level will rise by

3 feet. With luck—and a prodigious human effort at scaling back green-house gases—the rock stops there. But to date there is no evidence that humans are succeeding in any slowdown or reversal of the rates at which CO_2 and other greenhouse gases are being emitted into the sky and oceans. We seem ready to roll that boulder down an even steeper incline. We could end up with a 15- to 20-degree Fahrenheit temperature change and perhaps a 24-foot sea level rise that could ultimately leave the oceans 240 feet higher than they are now. Our greatest worry should be not only that the boulder may not need much of a push, but that once it is rolling down the steep slope there will be no stopping it.

LEARNING FROM CO_2 LEVELS IN THE NEAR PAST

We humans can very accurately keep track of carbon dioxide in the at-mosphere, and we can even measure the amounts that were present on Earth for well over the past million years, through careful analysis of minute trapped gas bubbles in ice cores taken from Greenland and Antarctic continental ice sheets. One of the most important, the Vostok ice core from Antarctica, has yielded a detailed record of carbon dioxide over the past 400,000 years. It shows that carbon dioxide varied between a minimum of 180 ppm and a maximum of 280 ppm.[14] Thus, for nearly 2 million years, atmospheric carbon dioxide values (and methane values as well, which mirror those of CO_2) went up and down, which led global temperature to seesaw as well. If we break down CO_2 levels into either above or below an arbitrarily selected level of about 240 ppm, it turns out that levels in the lower half of the cycle occurred more frequently than those in the higher half. During the low CO_2 times, the earth accumu-lated the great ice sheets—and we had ice ages.

Our planet did not break out of the 180–280 ppm range until about 1800, when carbon dioxide levels began to rise well beyond the old up-per limit.[15] By 1900, the level was 295 ppm, a rise of about 15 ppm over a century. But that was just the warm-up, so to speak. From 1900 to 2000, CO_2 levels went from 295 all the way up to the current level of about 385—a 90 ppm rise in just a hundred years. The curve these data describe will soon get even steeper. The rise will continue as China and India join

Europe and the Americas in putting two cars in every garage and heating millions of new houses with natural gas and oil. Even if carbon dioxide levels rose just another 90 ppm over the next century, by 2100 the atmosphere would have a CO_2 level of about 460 ppm. But most atmospheric scientists calculate future CO_2 levels by using the rate of rise over the past fifty years, rather the past hundred. Using those rates, which work out to about 120 ppm per century, we might expect CO_2 to hit 500–600 ppm by the year 2100. That would be the same carbon dioxide levels that were most recently present sometime in the past 40 million years—in terms of the rise of the seas, it would be equivalent to when there were little or no ice sheets even at the poles.

Yet climatologists assessing these new data dismiss even that scenario as too moderate. The rate at which carbon dioxide is increasing into the atmosphere is accelerating. Models using the latest values of the measured rise for the past decade, and projecting forward, lead to an estimate that CO_2 levels will nearly double in the next two centuries. By 2200, we might expect to see CO_2 levels approaching 1,200 ppm. Sooner than that, in as little as a century, levels might approach 1,000 ppm. That is the level of the Mesozoic Period and will cause the ice sheets to rapidly melt—all of them.

CARBON DIOXIDE AND OCEAN ACIDIFICATION

Not only do greenhouse gases destroy ice caps, but they themselves are lethal as well. The activity of these gases directly kills by carbon-dioxide or methane toxicity. To this we can add another potentially lethal process: acidification. To understand it, we have to digress briefly into ocean chemistry.[16]

Carbon dioxide participates in reactions with many other molecules. Several of these reactions are directly involved in maintaining the acidity or alkalinity of the ocean. The chemical species bicarbonate (HCO_3) forms part of the chemical buffer system that maintains a relatively neutral level of the oceans, making them neither acid nor base. However, if atmospheric CO_2 rises, the ocean becomes more acidic through a chemical reaction leading to formation of hydrogen (H+) ions in the sea. We measure the concentration of this H+ level using the pH scale, with lower values corresponding to higher acid levels. At small levels, a rise in acidity poses no danger to organisms. But if the levels rise enough, organisms are directly threatened. Rising acidity is most dangerous to organisms that produce calcareous shells, such as coral reefs and a type of phytoplankton called coccolithophorids. Moreover, once the acid levels rise, they linger at high levels for a long time: an ocean pH change would persist for thousands of years. Because the rise caused by carbon dioxide in fossil fuel happens faster than natural CO_2 increases have in the past, the ocean will be acidified to a much greater extent than has occurred naturally in at least the past 800,000 years.

If through most of geological time the CO_2 level in the atmosphere was higher than now, does that mean the oceans were once more acidic? At least for the past 100 million years, this was probably not the case. If there is lots of calcium carbonate in the upper reaches of the ocean (as there is when there are abundant blooms of the organisms that make chalk, the coccolithophorids, or another group called foraminifera), the ocean is described as "buffered"—neutrality is maintained despite the high CO_2. But buffering takes time, and that is the biggest difference between the effect of today's rise in CO_2 compared to any time in the past. During slow natural changes, the carbon system in the oceans has time to interact with sediments and therefore stays approximately in steady state with them. For example, if the deep oceans start to become more acidic, some carbonate will be dissolved from sediments, a process that tends to buffer the chemistry of the seawater and lessen pH changes. But what humans are doing in terms of injecting carbon dioxide into the oceans from human-made emissions is unprecedented.

The present-day rise in CO_2 seems to eclipse any other past rate of rise. This rapid rise outstrips nature's buffering systems, resulting in ocean acidification. Past atmospheric concentrations probably would not have led to a significantly lower pH in the oceans. The fastest natural changes we are sure about are those occurring at the ends of the recent ice ages, when CO_2 rose about 80 ppm in the space of 6,000 years. That rate is about one-hundredth of the changes we are witnessing now.

HOW GREENHOUSE GASES CHALLENGE PREDICTIONS
ABOUT NATURAL CLIMATE CHANGE

Climatologists have long theorized that climate change observed over the past 1.5 million years—long periods of very cold climate with growing ice sheets and dropping sea level alternating with shorter times of warmth—was the result of orbital changes.[17] All planets travel around the sun in an ellipse, not a true circle. Because of this, sometimes during the year Earth is significantly closer to the sun, and hence warmer, than other times, because when a planet is closer to the sun, more energy strikes it. Year in, year out, sometimes the planet is closer, sometimes farther away, as we make one complete orbit every 365 days. If this was all there was, understanding climate would be much easier. But the very elliptical orbit itself slowly changes position over thousands of years. Thus, over tens of thousands of years the earth's closest encounter with the sun takes place during summer or winter, or fall, and all in between. This leads to long-term variation in the actual amount of solar energy hitting all planets in our (or any) solar system. But then scientists developed the technology to drill into ice cores from Greenland and Antarctica to ascertain the temperature and carbon dioxide levels that occurred in the past. Their findings showed directly measureable changes in both carbon dioxide and global temperatures of very high precision. They found that there could be changes in global temperature in decades as well as millennia, and that every case correlated to CO_2 levels. The new finding was that both long-term and shorter-term variations in global temperature occurred and that these longer-term cycles helped alter glacial and interglacial episodes over the past 2.5 million years that together are termed the ice ages.

What might we have expected had humans not begun a grand global experiment in planetary engineering? One recent prediction was that the current "interglacial" period should end within a few more thousand years, to be followed by a drop of global temperature by as much as 20 degrees Fahrenheit for the next 80,000 years—spawning another cycle of continental ice sheets all the way from the poles to midlatitudes in both hemispheres. If this is so, in all that time our planet would never experience temperatures approaching present-day highs. But with global warming from increasing greenhouse gases, it is pretty certain that such a cooling has been put on hold. For how long, we cannot yet say, but surely for the next 10,000 years, if various estimates of increasing greenhouse gases and their residence time in the atmosphere and oceans are to be believed.[18]

PREDICTING FUTURE CO_2 RISE

One of the best sources of data about all aspects of climate is the award-winning Web site RealClimate.org. Run by climate experts without political axes to grind, this site has been a voice of reason and information for years. Much of this book is based on information published or referenced there. RealClimate.org provides two potential models of the future of carbon dioxide levels—and thus the future of our planet.

Barring the construction of a time machine, there is no way to know the actual arc of our planet's fate in the future. The climate we will have, the temperatures in a given year, the height of the sea—all are exactly unknowable. But at the same time, ever more sophisticated computer models, using values of known climate "forcings" (such environmental factors as the amount of energy from the sun as well as the level of greenhouse gases that "force" the climate to change), continue to yield new believable insights because they have been corroborated from modeling the past. When it comes to assessing how much carbon dioxide will rise, a 2006 RealClimate.org column by Malte Meinshausen, Reto Knutti, and Dave Frame offers a useful discussion and summary of the problem.[19]

The three authors go through the math showing that a stable CO_2 level of 400 ppm compared to our 380 ppm will yield an 80 percent chance that the earth will warm about 4 degrees Fahrenheit. For instance, the rise from CO_2 levels of 280 ppm at the start of the Industrial Revolution to the present 380 ppm has brought about a 1.5 degrees Fahrenheit global temperature rise, thus calibrating the climate models used to predict future temperature rises tied to rises in greenhouse gas concentration. The good news is that methane, one of the most troublesome of greenhouse gases produced by human activity, has a short life in the atmosphere before it breaks down. Also, the oceans are an effective sink for atmospheric carbon if we can sharply curtail emissions. Then concentrations of all greenhouse gases could begin to decline near the end of the twenty-first century (according to the best models now available, which, however, are just that—models). The three authors' model even lets greenhouse gas levels peak at 475 ppm for a short time, but if we brought them back down to 400 ppm before the end of the century, temperature rise would stop at 4 degrees Fahrenheit. Based on ever-newer reports of the rise of CO_2, however, that goal seems ever more out of reach.

There are other models of the potential greenhouse effect, the most important coming from NASA climate scientist James Hansen and his colleagues.[20] Hansen believed that one of the most powerful means by which the earth naturally removes CO_2 from the atmosphere is the natural dissolution of the gas into the ocean. So if a good model is to be made, the rate of this must be known, as well as the other variables that control global heat, such as the amount of ice on the planet at any given time (ice is highly reflective and thus cools the planet). Unfortunately, Hansen believed, these variables were not measured well in the past, so his group tried a more complicated method, not only including the amount of CO_2 in the atmosphere and the amount of sunlight being absorbed by the earth rather than reflected back into space, but also looking at the amount of water vapor in the atmosphere at the time being modeled. They added several other peculiar parameters, such as the amount of black carbon, organic carbon, ozone (an oxygen molecule that protects us from cosmic rays from outer space), the amount of H_2O high in the atmosphere as the result of methane breakdown, and aerosols also high in the atmosphere, especially molecules with sulfur attached, as well as other aerosols separately fed into the model to account for amounts of nitrates. Hansen also factored in oceanographic values, including temperature of deep and shallow seas and velocities of major ocean currents. The values of these variables are known from the past, and by including them, Hansen gave us a highly efficacious model. He determined the approximate global temperature among a host of other atmospheric variables for a series of past years, and thus could give realistic estimates of future global temperatures as well as oceanographic and atmospheric conditions. Further, the model compares the various forcings and ranks them in relative importance.

At the moment, the greatest uncertainties among the forcings come from the effects of the various natural CO_2 reservoirs in the ocean, places where gaseous CO_2 dissolves into seawater, and especially their rates of CO2 uptake. Currently the oceans are our best allies in scrubbing CO_2 out of the atmosphere, removing an enormous mass of the gas every year. Unfortunately, as we have seen, the net effect is to make the sea increasingly acidic so that the coldest places are the most acidic. Already researchers following the coldest areas report that the water's rising acidity is having biological consequences—most noticeably on the small planktonic creatures that produce calcareous shells. The increasing acid interferes with this calcification process and in the worst case causes the shells to dissolve. Because cold water absorbs more CO_2 (and oxygen) than warmer water, the global warming of the oceans will cause this sink to reduce. That, in turn, will increase the amount of CO_2 entering the atmosphere rather than the oceans.

Another major concern for Hansen and others in assessing the forcing factor comes from the vast amounts of methane and CO_2 held in permafrost and in frozen methane blocks known as clathrate. Methane is an even more potent greenhouse gas than CO_2, and in the atmosphere it converts to CO_2 over only a few decades. But prior to its decay to the lighter gas CO_2, it traps heat in the atmosphere very efficiently. The danger here is that rising temperature will increasingly liberate these currently inert forms of methane and CO_2 into gaseous form, thereby increasing atmospheric levels and greenhouse effects.

The bottom line from all this work is that the rate of global warming is far higher now than the "consensus" levels from other models. If correct, Hansen's findings indicate that sea level will rise ever more rapidly.

Because the earth's heat budget is complicated by the effects of the oceans, land, and especially air and water currents, CO_2 rise and global temperature do not share a linear relationship. The rule of thumb climatologists use is that each doubling of the CO_2 level can be expected to raise global temperatures by about 4 degrees Fahrenheit. Thus the projected CO_2 level even for a century from now would raise the global temperature between 4 and 7 degrees. Today that temperature is estimated at between 59 and 61 degrees Fahrenheit. This does not mean that everywhere on Earth has this temperature—but that the planet taken as a whole does. It would climb to as much as 68 degrees Fahrenheit. The effect of that escalation would be earth changing and could conceivably bring about the greatest mass death of humans in all of history.

Despite the various factors being considered by climate scientists such as Hansen and the RealClimate.org researchers, both climate-change professionals and the public focus primarily on the actual amounts of CO_2 currently being emitted into the atmosphere. Thus, a September 2008 report on this figure came as a bit of a shock. Rather than showing lowered values of CO_2 entering the atmosphere, 2008 promised to have the highest rate ever. It was expected that the economic downturn of 2007 and 2008, coupled with efforts to lower emissions, would have had some effect. But the published figures for 2007 showed a 3 percent increase in the amount of carbon put into the atmosphere compared to the year before. This figure exceeded the worst-case scenario of the Nobel Prize–winning group of climate scientists who have been avidly monitoring the effects of carbon emissions.

National governments have indeed paid at least passing attention to reducing emissions, even as their larger efforts to lessen climate change have been thwarted by short-term concerns and internecine bickering. Concerted efforts to reduce emissions are well under way, but working against these attempts has been the continued industrialization of the world, and especially the phenomenal growth of India and China. Both countries need vast new sources of power, and coal is cheap compared to oil. Thus, at the start of the twenty-first century, new coal-fired plants have become a major source of CO_2 emissions, and greenhouse gases have increased.

The leader in carbon emissions for 2008 was China, followed by the United States. While several countries did succeed in lowering emissions, the United States produced more. The total came to 9.34 billion tons of carbon put into the atmosphere globally. As we will see in a chapter to come, that figure will soon be dwarfed by even greater industrialization and consequent carbon emission. And the sea will rise.

World energy use is the prime cause of anthropogenic emissions of carbon dioxide. Whether these emissions rise or fall will depend on the world economy as much as world conservation. The belief that the future of the atmosphere is really in the hands of the Chinese and Indians has some truth to it, but it would be foolish to exculpate North America and Europe from responsibility for curbing carbon emissions. The latest projections have emissions rising from 28.1 billion metric tons in 2005 to 34.3 billion metric tons in 2015 and 42.3 billion metric tons in 2030. Not good news.[21]

FAILURES OF INTERNATIONAL ACCORDS

No book about sea level rise would be complete without some mention of the colossal international failure of American leadership known as the Kyoto Protocol, the 2005 international accord that requires some major industrialized countries to reduce their greenhouse gas emissions collectively from 2008 to 2012 to an annual average of about 5 percent below their 1990 level. The United States and other powerful nations hamstrung the agreement from the start, by allowing it to exempt fourteen countries considered "economies in transition" including China and India, which both ratified the treaty merely as a public relations move. The United States refused even to ratify it.

Kyoto has had at least the effect of bringing more attention to the problem globally—but seemingly no effect on slowing the rise of CO_2.[22] World energy-related carbon dioxide emissions are projected to grow by an average of 1.7 percent per year from 2005 to 2030.[23] The highest rates of increase in annual emissions of carbon dioxide among nations are China at 3.3 percent, India and other Asian countries at 2.6 percent, Brazil at 2.3 percent, the Middle East and Africa at about 1.8 to 1.9 percent each, and Russia at 0.9 percent. In contrast, the United States is projected to increase emissions annually at 0.5 percent. Although that number is relatively low, the sheer magnitude of the U.S. economy, largely powered by carbon energy, still ranks it among the lead emitters, now and well into the future.

HOW MUCH IS TOO MUCH CO_2?

As we confront the reality that greenhouse gases are radically altering our atmosphere and world climate, we have to ask whether increased temperatures will reach either levels high enough to cause positive feedback that incites even faster rises, or levels that are irreversible in the near future. Yet the actual level where these various and potentially catastrophic events take place is highly disputed and will remain so until some kind of tipping point occurs.

The controversy over a potential tipping point has not stopped various climate scientists from hypothesizing about what could irreversibly happen when.[24] Of these researchers, most noted is James Hansen (introduced earlier in this chapter), who is viewed by his detractors as Chicken Little and by his supporters as Jeremiah—and who also was a victim of Bush administration censorship on climate change. In 2008, Hansen and an astonishingly large assemblage of forty-six coauthors used various climate models to look beyond the simple estimates of rise from their previous work. This time they specifically looked for temperature and CO_2 levels that might prove to be catastrophic in that it would be too late to put the genie back in the bottle—or in this case, the CO_2 back into rocks. They called this catastrophic level a "tipping point."[25]

For some of the same reasons that make weather forecasting so difficult, the complicated physics of a world ocean and atmosphere constantly in motion means that any model for a tipping point will evince a significant range of results. A further complication is that the climate system, made up of the various components that cause weather and global temperature and its changes, possesses what has been termed "thermal inertia"—which means there is a lag time of one to several decades before the climate responds to changes in one of its drivers, such as rising greenhouse gases in the atmosphere. An even more serious complication is that while any such model must look at the aspects of physical changes over decades, the human component is also changing at the same time. For instance, the rapid construction of coal-fired power plants in China and the increase in the number of cars worldwide, to name but two examples, have import for our future.

In an attempt to answer how much carbon dioxide is too much, Hansen summarized the group's findings as follows:

The Intergovernmental Panel on Climate Change and others used several "Reasons for concern" to estimate that global warming of more than 2–3°C may be dangerous. The European Union adopted 2°C above pre-industrial global temperature as a goal to limit human-made warming. Hansen et al. argued for a limit of 1°C global warming (relative to 2000, 1.7°C relative to pre-industrial time), aiming to avoid practically irreversible ice sheet and species loss. This 1°C limit . . . implies maximum CO_2 ~ 450 ppm. Our current analysis suggests that humanity must aim for an even lower level of Green House Gases. If humanity wishes to preserve a planet similar to that on which civilization developed and to which life on Earth is adapted, paleoclimate evidence and ongoing climate change suggest that CO_2 will need to be reduced from its current 385 ppm to at most 350 ppm. The largest uncertainty in the target arises from possible changes of non-CO_2 forcings. An initial 350 ppm CO_2 target may be achievable by phasing out coal use except where CO_2 is captured and adopting agricultural and forestry practices that sequester carbon. If the present overshoot of this target CO_2 is not brief, there is a possibility of seeding irreversible catastrophic effects.[26]

In this chapter we've established that the atmospheric concentration of carbon dioxide has an enormous effect on global temperatures. Unfortunately, many climate skeptics still refuse to believe this. Yet the evidence is there—and perhaps none is more compelling than in the rock record's evidence of warm periods and cold periods coinciding with, respectively, CO_2 levels higher or lower than today.

Carbon dioxide affects our lives in many ways, a number of them trivial to most of humanity—such as one I discovered over a very slow and satisfying lunch with vintner Robert Mondavi in California's Napa Valley a few years ago: if the climate warms, what will happen to the distribution of the French oak, whose wood is needed to make the best of Mondavi's cabernet sauvignon? But others, of course, are vital to our survival. How quickly will food crops be affected by climate changes in a way similar to the loss of the French oaks? What will warmer winters mean to apple production when the necessary chill producing dormancy does not happen, or when the heat level becomes too high for cattle to reproduce? And perhaps most important, how high will sea level rise? Perhaps none of the myriad estimates of climate change is more uncertain than what the seas will do. Calculations of the increase in sea level are subject to vari-ability between minimum and maximum estimates of rise, between the earliest and latest times that the sea could reach a certain level, and even in estimates of how high sea level will rise before tapering off. These are the projections that have the greatest bearing on future human history, estimates that range from minor annoyances to death sentences for some proportion of humanity.

CHAPTER 3

THE FLOOD OF HUMANS

El Kef, Tunisia, 2060 CE. Carbon dioxide at 500 ppm.

The call to prayer jerked the old geologist out of his troubled sleep in the predawn darkness of his hotel room. A gasping Toyota pickup, belching fumes in the gloaming, roared past the dilapidated hotel with speakers like an ancient megaphone to amplify the grating call, sending their tinny sonic message into every house, from the grand mansions down to the modest hovels clinging to the edges of the town—time for first prayer.

He rolled over in the creaking bed, groaned again at the knowledge that there was no hot water in the dripping shower, and dressed in the near dark. Since he was not a Muslim, it was not the mosque that was his goal, but the huge white hills behind the town, great sections of limestone and marl that held a record of the deep past, when this entire region was several hundred feet beneath a sea that was much higher. This geological section recorded a past mass extinction caused by a short-term rise in carbon dioxide that looked eerily like the increase occurring in his own time. By this time in the twenty-first century, CO_2 was over 500 ppm, and the mass extinction seemed to have happened when CO_2 levels shot up to 1,100 ppm. The difference was that in the past the extinction was caused by volcanoes, not Volvos—and not other cars, power plants, and airplanes, among so much else.

This part of Tunisia had not seen the westernization that for generations had made Tunis, the capital and largest city in the country, a beacon of tolerance and modernization for all of the northern African shore. In many ways it was as it had been in the Middle Ages, and the geologist's skin still tingled from yesterday's appointment in the misnamed Turkish Bath, a stony cave where thick steam and a thousand years of male sweat had infused the old stones with the rank odor of humanity, and where the muscled masseur had scraped his naked skin with pumice stones following the brutal kneading that passed as a massage here. For most of the fellow bathers this was the weekly cleansing of the body that accompanied the several-times-a-day cleansing of the mind in the ancient mosque. As its population of believers continued to swell, Islam grew ever more conservative, thanks in no small part to the drought affecting all of North Africa as well as the Middle East. As the countryside withered under a decade of thirst, the most dire in living memory, the people's only hope was that God would intervene.

Breakfast was bread and strong coffee, and he put dates, olives, and two more small loaves of bread in his knapsack for his midday meal—provisions that would cost a relative fortune for the people here. As usual he sat as far from the hotel's windows as possible: the hungry masses in the street, thronging one of the few places where westerners could get food on any day, troubled him to the point that he could no longer simply say no, or give away all his food, as he had during his first few days here. In North Africa, hunger was almost visible. Tunisia, once the granary of the Roman Empire and as recently as the late twentieth century easily able to feed itself, was starving, as were the other fresh water–challenged states of Algeria, Morocco, Libya, and Egypt. Only the largesse of the still-rich oil sheikdoms kept the famine from getting worse along the North African shore. Almost paradoxically, the onetime sinkholes of famine—Ethiopia, Sudan, and Chad—were not as badly off, because the endless wars they fought over tribalism and religion had abated as new rains, perversely initiated by new atmospheric wind patterns, brought grain and livestock to their once empty lands to the south of the Mediterranean states. But the northern rim of Africa was still without industry that would at least allow its ever-burgeoning populace to purchase food. There was no way the countryside could sustain the hundreds of millions. Many starved.

The geologist set off on foot as the sun broke over the valley. Passing through town toward the day's work, he was soon hiking into the hills, which was in no way a passage into a human-free region. The hills sprang from withered fields, acreage that had been verdant and productive on his previous trips here when he was a young man, at the end of the last century. Now all was brown. Around the fields were the small stone houses that characterized this region, each one featuring a single wire for the electricity that had arrived here only some forty years earlier. Around these houses a veritable flock of children played in the early morning, the younger ones gathering in groups, the older boys already prowling as packs. When last here he had been surprised at how many children there were, and now he was undoubtedly seeing the children of those past kids, the numbers now multiplied by four.

He began his work, the slow, meticulous job of measuring and describing the thick, white, Cretaceous-age limestone, collecting the odd fossil as he did so. As always, he was soon surrounded by a mob of kids. Some wanted to crack his hammer on nearby rocks, or nearby heads; others wanted to take their own measurements. Soon there were at least fifty boys around, with no girls to be seen at all. The boys came from the huts around these foothills, and El Kef proper was also awash in children. Their presence in such great numbers was no surprise, because in this year the world had achieved a dubious distinction—for the first time it was the home of 9 billion humans.

The North African shore had certainly made its own significant contribution to that so-called milestone. After all, North Africa had the highest birthrate on the planet, and with little AIDS here, there was none of the check on overpopulation that the devastating epidemic still exacted from sub-Saharan Africa. The modicum of health care provided here, and the accompanying reduction in infant mortality, had turned Tunisia, and nearby Algeria and Morocco, into human-producing factories.

North Africa was not alone in possessing the demographic and medical conditions, and the social and religious mores, that caused great population growth. Human populations were swelling like never before in many parts of the globe, and in an economically perverse way—the richer the country, the fewer the babies, as more and more women held off having children while they completed college or other forms of postsecondary education. Because of this, the rich countries had slower population growth, or even, in the case of Eastern Europe, Germany, and Japan, negative growth. But the surge of nationalism in those regions had closed many doors to the immigration that had previously siphoned off a good deal of the overpopulation in Turkey and the northern African states. Holland had finally closed its doors to Muslim immigrants, following the lead of Germany, France, England, and Spain. The United States had its own problems. Its southern half was now dominated by the descendants of immigrants from Mexico who multiplied more plentifully than their Anglo neighbors, exacerbating ethnic tensions and contributing to America's own near doubling of population to 500 million people and the recent bankruptcy of the Social Security and Medicare programs. The nation was facing its own dark age.

The old geologist tried to gently shoo away the more persistent kids so he could get back to work. Some of the children were easier to ignore, for they were skinny sacks of bones, with little energy. He had taken to starting his fieldwork early and quitting well before the blistering noonday sun would make all work impossible. Soon he would be ending this project, and it was with great grief but no regret that he would leave this sea of children in a country with no prospects, no water, and little food as the African drought wore on. He was as much a fossil as those he collected: the world had no need for paleontologists. Learning for learning's sake was a luxury no society could afford.[1]

THE HUMAN FACTOR

In the last chapter we showed that in the absence of human intervention, atmospheric carbon dioxide does rise and fall over time, due to exchanges of carbon among the biosphere, atmosphere, and ocean, and, on the very longest timescales, what is called the lithosphere (rocks, oil reservoirs, coal, carbonate rocks). The rates of those exchanges are now being completely overwhelmed by the rate at which we are extracting carbon from the lithosphere and converting it to atmospheric carbon dioxide. One profound driver of CO_2 escalation is the ever-increasing human population. In this chapter, we will look at the phenomenon of population increase and its direct side effect, the increase in energy being transformed from some reduced chemical compound (one that can react with oxygen to "burn" and produce carbon dioxide as a by-product) and the rise in sea level. Furthermore, if it can be shown that human population increase has caused rising atmospheric carbon dioxide levels—but that humans can undo their damage—then it is important to recognize how long-lasting the current and near future carbon dioxide levels will be.

The best estimate comes from University of Chicago oceanographer David Archer, who in my opinion has done more than anyone to show how long CO_2 created by fossil fuels will persist in the atmosphere. For example, in his book *The Long Thaw* he noted: "The lifetime of fossil fuel CO_2 in the atmosphere is a few centuries, plus 25 percent that lasts essentially forever. . . . The climatic impacts of releasing fossil fuel CO_2 to the atmosphere will last longer than Stonehenge, longer than time capsules, longer than nuclear waste, far longer than the age of human civilization so far."[2] Because it appears that fossil fuel CO_2 can persist in our atmosphere even longer than the 50,000 years radioactive material remains toxic, we need to take a hard look at the effect caused by the very fact of our species' burgeoning numbers. More fossil fuels are consumed by more people, and the planet can bear only so many billion carbon footprints.

As I give talks around the country about a newly discovered phenomenon of the deep past greenhouse extinctions, people always ask about the relevance of these studies to the present and near future. That question is simple to answer, at least for me: what happened in the past can and will happen again if we continue to heat the planet at present rates. But another question they ask involves something quite different than extinction. If the increasing number of humans affects planetary temperature, then why is virtually no one talking seriously about some form of population control? Interestingly, the people usually asking this question are almost universally beyond childbearing age; in fact, most people coming to my public talks about climate change are no longer having children. I think I know the genesis of their query. Back in the 1970s there was a powerful group calling itself Zero Population Growth—or ZPG, an acronym that readily stuck in the cultural consciousness.[3] But when the movement ran smack into the Republican takeover of the United States in 1980 and the issue of overpopulation faded, ZPG's influence waned. Yet all the while the number of humans kept increasing. Every new human mouth to feed, or transport, or simply to keep alive, warms the planet, which melts the ice, which causes the sea to rise.

A lot of humans are currently on Earth—in 2008, that number hit 6.7 billion.[4] Human population has grown exponentially since the 1950s. The highest growth rate of all time seems to have occurred in 1963, according to those keeping track for the United Nations, when the population increased by 2.2 percent largely through the great number of births in Africa and Asia, where as many as six children per childbearing female was not uncommon. Although the rate of growth as a percentage of the population has halved since then, in absolute numbers more humans are on the planet every year. The rate of growth is such that human population will reach 9 billion in 2042. And it does not stop there. As in sea level rise, where most forecasts look to 2100 and then stop (enabling the inaccurate assumption that the rise will just stop that year as well), human population forecasts seem to look at that 9 billion mark as the end of things. But that is far from true. By 2150 population is expected to reach nearly 9.8 billion, and it might keep going up, although it looks like by that time the juggernaut of population increase will slow. But the prospect of "slowing" points to only the bright side of human population predictions. The reverse of this relatively good news (as if having only 9.8 billion people is good news) is that some estimates suggest that in the future as many as 30 billion people might occupy the planet at one time. I believe that, more than any other factor, the ever-increasing number of humans causes the seas to rise. The modern food production and energy they require emit greenhouse gases. So far, rising human population is highly correlated to rising emission rates—and to the absolute level of CO_2 in our planet's atmosphere.

It is agreed pretty much worldwide that reducing emissions is wise. But one way, perhaps the only way of doing so effectively, is to lower human population. Barring a devastating famine, a pandemic that kills hundreds of millions, an asteroid impact, or a catastrophic war, a reduction in population is unlikely in this century—and as we shall see in this chapter, it may not happen in the next century either.

PREDICTING POPULATION

Every decade the United States conducts a census. This is a pretty important enterprise, because much of the nation's planning and economic policy, region by region, is based on the results. Unfortunately, many people do not cooperate in this endeavor or are skipped inadvertently. Thus, even in the industrialized, well-wired United States, taking the census is never smooth, and the results can only ever be approximations, with undercounts inevitable. But if things are a bit problematic for measuring U.S. population levels, the situation is far worse for the world as a whole, especially in nations that do not have the economic or social stability required to count their people. Ascertaining the real number of humans on our planet is impossible, and we must rely on good guesswork.

The United Nations has long been entrusted with conducting the world census, producing voluminous volumes showing who and what areas are increasing in population (and the few that are decreasing, too). But a second agency also conducts its own census and makes its own policy statements based on the figures: the American Central Intelligence Agency. Analysts at the CIA have long recognized that population rise in many parts of the world is a national security issue—a threat to the nation's long-term stability.[5] Both reports hypothesize that there will be unequal, in some cases highly unequal, changes in regional population as the years go forward. The absolute numbers as of July 2009 are shown in Table 3.1 on the following page.

The European figure is alarming to Europeans. Not only are they *not* replacing themselves but they also are failing to reproduce at a rate that would maintain their current population.

Is there any robust prediction of global human population after 2100? It seems that only the UN is making estimates beyond that year. In 2004, the Population Division of the UN Department of Economic and Social Affairs prepared a set of population projections to the year 2300 for each country.[6] Those estimates are already dated, but they are the best we have. All projected scenarios share the same assumptions about the steady decline of mortality after 2050 and the consequent increase of life expectancy. Thus some of the population increase will be from fewer deaths as a fraction of the population as well as the arrival, and survival, of more children.

TABLE 3.1. World population and average babies per mother per region

Region	Current Human Population	Number of Children per Mother
All Asia	4,052,000,000	2.5
India–Pakistan-Bangladesh	2,450,000,000	3.0
China	1,324,700,000	1.6
Latin America and Caribbean	577,000,000	2.5
Northern America	915,000,000	2.1
Oceania	36,000,000	0.5
Europe	736,000,000	1.5
Northern Africa	195,000,000	4.5
Central Africa	138,000,000	6.1
Southern Africa	650,000,000	4.8
All Africa	973,000,000	5.0
World	6,707,000,000	2.6

The UN sees three possibilities: a low, medium, and high level of population growth. All three project a slowing of population growth following the high-tide mark of 9 billion in 2040. The low estimate is in fact quite encouraging: it predicts a decline in population to less than today's levels by 2100, and thereafter a steady reduction in numbers to a level of about 3 billion in 2300. Thus the rosy view.

The second set of estimates is the middle road, and it projects a very different scenario. In this projection, population actually stays at the 9 billion mark, even reaching 9.2 billion as late as 2075. It thereafter declines slowly, but there are still 8.3 billion people on the planet in 2175. But after this time, the decline is over. Population jumps upward again, hitting the 9 billion mark a second time, in 2300, which is as far out as this study goes.

Finally, there is a scenario of unending population growth. It is a prospect that is shocking even to contemplate. In this projection, human population takes a brief respite from its upward growth in 2075, but then it shoots up again and keeps on rocketing higher, with world population hitting an estimated 20 billion people by 2200, and an astonishing 37 billion in 2300. But it would not stop there, as the trend near the end grows even steeper—50 billion in 2400, anyone?

The lack of certainty in any of these projections frustrates those trying to plan for what could happen to the level of human population and to the accompanying expenditure of resources that a crowded world would require. Because population numbers are so tightly linked to resource use—and to the resulting emission of greenhouse gases—any rational planning or even warning of impending environmental peril comes with vast uncertainty. As an example of the unpredictability of human reproductive behavior, China's draconian efforts in the past century to limit each woman to just one child worked very well, to the surprise of many who study human behavior. It is sufficient testimony to the ability of nations to control their own populations in the face of dire circumstances. But will that injunction be adopted by any society in the near future? The Chinese experiment may have worked because it was imposed on a cowed people by a totalitarian state. China's citizens might not be so docile a second time—and it may be impossible for other societies to take command of the human imperative to procreate.

THE HUNGER FOR ENERGY

All people use energy, from not much per person in sub-Saharan Africa to too much per person in Europe and North America—too much, that is, if we want a sea level outcome different from the one foreseen in this book. A world with more people will use more energy and create more greenhouse gases—an effect further aggravated by the westernization of India, China, and the rest of Asia, with every family's goal not just a chicken in every pot but a car in every garage. Only a revolutionary change in personal, corporate, and public use of energy could avert a disastrous increase in atmospheric emissions.

Assuming that energy use continues along the path it took in the first decade of the new century, predictions can be made about our energy requirements in the near future as the number of humans on the planet inexorably grows. The U.S. Energy Information Administration (EIA), which is made up of not just American experts but engineers and economists from many different countries, tracks energy use in its annual *International Energy Outlook*, a report that might be the single best crystal ball for the future energy needs of the growing global population.[7] The EIA report divides world countries into either industrialized or non-industrialized nations. It estimates that energy needs for the industrialized countries are expected to increase 25 percent from 2005 values by 2030—while energy for the non-industrialized nations (where human population growth will be greatest) will essentially double. Taking into account the actual energy consumption of these two groups of countries, it is estimated that in 2030 worldwide demand for energy will be 50 percent more than the world produces today, creating an ever-greater demand for energy.

Because producing energy leads to the production of emissions, the EIA also estimated the rise in CO_2 over the same period. The EIA predicts that world carbon dioxide emissions will increase steadily from the 28.1 billion metric tons emitted in 2005 to 34.3 billion metric tons in 2015, and 42.3 billion metric tons in 2030. However, the increases will not be uniform across the globe. The so-called developing countries will show the greatest increases in generated carbon dioxide. By 2030, emissions from developing countries are projected to exceed those of the current industrial powers (the United States, Canada, Europe, and parts of Asia) by about 70 percent. That means a huge addition in the amount of CO_2 going into the atmosphere in the near future. It also means we will need a lot of new energy powerhouses to fuel the worldwide pursuit of prosperity.

Where will all this new energy come from? The current answer seems to be coal—of all energy sources, the single highest emitter of greenhouse gases.[8] World coal consumption is projected to increase from 123 quadrillion Btu in 2005 to 202 quadrillion Btu in 2030, at an average annual rate of 2 percent. Coal's share of world energy production has increased sharply over the past few years, largely because of strong increases in coal use in China (which has nearly doubled its coal burning since 2000). China appears poised to increase its coal burning to even higher levels and alone will account for about 70 percent of the future increase in world coal consumption. The United States and India—both of which also have extensive domestic coal resources—each account for only 9 percent of the projected world increase in coal burning. But that seemingly low number belies a great deal of new carbon dioxide going into the atmosphere.

In the American presidential election of 2008, there was a vigorous debate about the nation's projected domestic energy use, much of it dealing with the question of when or if the United States can become energy "self-sufficient," a situation that would be brought about by ceasing oil imports, largely from the Middle East. But because the United States has already extracted a great deal of its oil reserves, it will have to tap other energy sources. Both candidates, Barack Obama and John McCain, touted the development of non-polluting solar, wind, and tide sources of energy, and there was much talk about recommitting to nuclear power. But when all the electoral dust settled, it became clear that the greatest source of energy would remain coal.

THE END OF OIL

Oil is a finite resource, and with the new century came increasing reports that the world was either entering, or perhaps even concluding, the period of "peak oil"—the time when a maximum amount was extracted worldwide—to be followed by an irreversible decline exactly when world population would explode, with its inevitable demand for more oil. Peak oil is also called "Hubbert's Peak, named for the Shell geologist Marion King Hubbert, who, in 1956, accurately predicted that US domestic oil production would peak in 1970."[9] He also predicted that global production would peak around the year 2000, which it would have, had not the politically created oil shocks of the 1970s delayed that milestone for five to ten years. According to the Hubbert model, the production rate of a limited resource will follow a roughly symmetrical bell-shaped curve. Some figures now suggest that peak oil occurred in 2005.

How much is left? The amount of conventional oil-in-place is somewhere between 6 trillion and 8 trillion barrels, though this number has grown over time with discovery of new sources, such as the Brazilian oil fields, estimated to be fifty billion barrels alone, and the fact that countries such as China and Iran may be withholding information about actual reserves for national security reasons. The volume of non-conventional oil-in-place, such as oil shale and tar sands, is rather murky, with an estimate of between 7 trillion and 8 trillion barrels. Taken together, there may be between 13 trillion and 16 trillion barrels of oil left from all sources. Note, however, that this number represents the total oil existing on our planet—not the total that is recoverable. That figure is much smaller, an estimate that constantly changes because of two factors: new technology and practices, which will enlarge the number, and what appears to be a serial overestimation of currently producing fields, which deflates it. The bottom line seems to be that only 35 percent of the oil in the ground is recoverable, and this is the optimistic view.

The world in 2009 used about 245 million barrels of oil per day; by 2030 that figure is hypothesized to increase to 325 million.[10] This enormous demand for oil to fuel high technology and a high-population human civilization will certainly tax our ability to feed everyone (since modern farming runs on oil-powered machines and uses oil-produced fertilizer), and it will strain our capacity to raise living standards in less developed nations while maintaining current living standards in the developed countries.

Governments, including the American government, are certainly taking notice. The American response typically was talk, not action. The Hirsch Report, created at the request of the U.S. Department of Energy and published in February 2005, examined the likelihood of the occurrence of peak oil, the necessary mitigating actions, and the probable impacts based on the timeliness of those actions.[11] The report's lead author, Robert Hirsch, published a brief summary in October 2005 for the Atlantic Council. I can paraphrase his conclusions as follows:

- World oil peaking is going to happen and will likely be abrupt.
- Oil peaking will adversely affect global economies, particularly those most dependent on oil.
- Oil peaking presents a unique challenge ("it will be abrupt and revolutionary").
- Mitigation efforts will require substantial time. Here are three possible scenarios:
    - A twenty-year transition to other energy sources; if other such sources are available, there will be little impact on world economies.

–A ten-year rush transition that would cause moderate impacts, which would create shortages but is still possible with extraordinary efforts from governments, industry, and consumers.
–A sudden transition, or late initiation of mitigation that might cause severe consequences.
- Both supply and demand will require attention.
- Government intervention will be required.
- Economic upheaval is not inevitable ("given enough lead-time, the problems are soluble with existing technologies"), although a betting man might not bet against a great upheaval at the end of peak oil.

These conclusions were followed by action scenarios that included undertaking a crash program to exploit alternative energy sources, either ten or twenty years before peak oil, or at the time of peak oil. But if peak oil has already come and gone, we may have missed the clarion call.

KING COAL

There are four kinds of coal, based on energy richness of each particular type.[12] The worst kind is lignite, followed by sub-bituminous coal, bituminous coal, and anthracite, from which the most energy can be extracted per unit weight. It is informally called hard coal, while the coal types falling toward the lignite end of the scale are soft coal. The softest coals are sometimes termed brown coal.

Coal has the reputation of being nearly inexhaustible—and seemingly by its sheer abundance, if for no other reason, the perfect substance to fuel a growing world population. But estimates for global coal resources have been revised down, by 55 percent over the past twenty-five years, from 10 trillion tons hce (hard coal equivalent) in 1980 to around 4.5 trillion tons hce in 2005.[13] Six countries (the United States, Russia, India, China, Australia, and South Africa) hold about 85 percent of world coal reserves. China, the world's largest producer of coal, has an unknown reserve—they might know but are not telling. The 2006 Statistical Review of World Energy estimated that China has fifty-five years of remaining reserves at current production rates (while depleting its reserves at almost 2 percent per annum). The United States has reported proven coal reserves that would allow continued production at current rates for more than two hundred years. However, many of these reserves are of low quality, with high sulfur content and other drawbacks. Measured in terms of produced tons per miner, U.S. productivity of coal steadily increased until 2000 but has declined since, which also implies that "easy coal" is running short. The United States passed peak production of anthracite (by far the rarest form) by 1950 and peaked in bituminous coal in 1990, but sub-bituminous coal more than made up for this decline in terms of tonnage. However, due to the lower energy content of softer coals, the total energy content of annual U.S. coal production peaked in 1998. Estimated maximum production from all U.S. coal types will probably coincide with what may well be the maximum world population, both occurring soon after 2050.

As we saw above, the International Energy Association (IEA) is in the forecasting business, and its latest estimate is that coal use will rise by 60 percent by 2030. Because most anticipated economic and population growth is expected to take place in so-called undeveloped countries, the majority of world coal use will happen there; even by 2030, it is estimated that more than 1.5 billion people still will lack access to electricity, and every population in every nation wants what the West has. For most of the world, producing electricity means burning coal. Nuclear plants, solar fields, and wind energy are all high-tech enterprises that might be beyond the current industrial capabilities of developing nations, and oil is too expensive for them. Thus, because their populations hunger for it, there is no conceivable political means through which that coal use will be rationed, or probably even monitored.

THE DUBIOUS PROMISE OF CLEAN COAL

Coal is the most polluting of all fuels—to the point that some prominent Australians are demanding that their country stop using and exporting it. However, that is not going to happen; in fact, coal is Australia's biggest earner of hard currency, with most of the country's coal exported to China. Coal makes up the majority of India and China's energy needs and will continue to, well into the future. Given coal's inevitability as an energy source, the world will need to figure out how to mitigate the effect on the world's ice and sea levels by somehow radically reducing emissions from coal-fired power plants. But is this even possible?

Coming from the coal-producing state of Illinois, President Barack Obama has publicly and repeatedly supported "clean coal" technology.[14] U.S. coal companies heavily promote it. It sounds great—clean coal must mean cleaner burning and less pollution. But just how do you "clean" coal—a black rock that just oozes pollution, from carbon dioxide, to nasty sulfur compounds with lead, to acid rain, upon burning?[15] Coal-cleaning technology does remove the ore's inherent impurities, especially the very dangerous sulfur compounds, to various degrees. But there is one massive caveat to clean coal technology: it does not reduce carbon dioxide emissions unless further modification and expensive equipment are added to the power plants. With clean coal, there may be less sulfur sifting itself into the sky, but there is just as much CO_2. In fact, in all existing plants, there will be as much CO_2 going into the atmosphere with clean coal as with dirty coal. The only hope for controlling coal-induced greenhouse gases is the construction of "gasification" facilities as part of new coal-fired plants, and the retrofitting of old plants. Gasification technology can remove or store a great deal of the CO_2, some of it in underground geological repositories. But it is very expensive technology, and the plants going on line currently in China and India do not have this technology as standard.[16] As more and more plants are built to serve the needs of an increasing and increasingly affluent population, the result portends disaster. Why does truly clean coal matter so much? Coal is nearly pure carbon, which means it releases nearly pure carbon dioxide when burned.

The smog in Beijing is legendary—and much of it comes from burning coal. But there will be far worse emissions wafting up from both India and China as more and more of their coal-fueled plants go on line. Yet China has offered one modest promise of progress. Voicing concern about the global and local environment, it has designed and commenced construction of a state-of-the-art coal power plant that would capture all CO_2 for sequestration in geological reservoirs. This welcome development is described as follows:

As a first step, a consortium of power and coal companies will fund the construction of an integrated gasification combined–cycle power plant, in which coal is converted into gas and the pollutants are removed before the gas is burned. The plant, in the port city of Tianjin, would produce 250 megawatts of electricity. Subsequent phases would boost output as high as 650 megawatts, a moderate size by utility standards. After that time, work would begin to capture CO_2 from the plant and sequester it in a depleted oil field nearby. Target date: 2015.[17]

In the aftermath of the U.S. government's decision to pull the plug on its own carbon capture project—FutureGen—China's $1-billion Green-Gen plant has become the world's leading effort on such technology.[18] But because China is suffering from coal shortages, and extra energy is required to gasify the coal and to capture the CO_2, completing GreenGen may prove a challenge. GreenGen must function as a for-profit power plant, so even though Chinese officials agree on the long-term benefit of removing CO_2 from the air, the future of climate change may depend on the economic value assigned to that benefit. And GreenGen is just one plant among a myriad in China, with the rest spewing carbon dioxide into the ever-warming air. The world climate's circumstances demand solutions, and instead we have one pilot program. We are left with the following dictum: reducing emissions from burning coal will be part of generating power only if it becomes cheaper than doing nothing with emissions but blowing them into the sky. That is a daunting, perhaps patently impossible task. If humanity cannot slake its thirst for cheap energy, the most important question is whether coal and oil will run out soon enough so that further catastrophic increases in greenhouse gases will be contained, the world's ice will not melt, and the rising seas will not flood our coasts. That question remains unanswered.

ENERGY, POPULATION, FOOD

Just as the increase in population will cause increased demand for energy, so too will the world be required to come up with more sustenance than we produce now. This subject is so important that a whole chapter of this book is devoted to it, Chapter 4. But because food supply requires oil and other energy sources (for fertilizer and mechanical aspects of planting and harvesting), it is relevant in this chapter to examine the relationship between food sources and energy output. Modern farming does indeed require an immense amount of petroleum not only for fertilizer but also for the technology that ensures high-yield harvests. Fuel is required for motorized farm and field equipment, and for the raw materials of fertilizers, because synthetic fertilizer—the most absolutely necessary ingredient in creating world food—is made from natural gas. In fact, right now about one-fifth of all U.S. energy use goes into the food system—from fertilizers to threshers to the many trucks and trains hauling food to market.[19] Some say that U.S. agriculture is completely dependent on oil—that it cannot convert to coal, for instance. And if indeed we have hit peak oil, the cost of producing food in a world where oil prices are skyrocketing will have profound effects. Much of the world today staves off famine because of the large annual U.S. agricultural product surpluses, which are distributed around the globe as a humanitarian measure. With high oil prices, those surpluses might disappear.

The rising cost of oil in 2008 (the year it rocketed up to $150 per barrel) drove home the lesson that the worst possible future scenarios are looking increasingly correct. High oil prices in 2007 and early 2008 had two effects that combined to raise the price of food staples radically. First was the initiation of large-scale biofuels production in many industrialized nations—including the biggest breadbasket of all, the United States—all of which sought to develop new energy sources from corn and other crops. Grain products, now in demand not only as food but also as fuel, saw large run-ups in prices. Even such "necessities" as beer (certainly an essential fuel for every geologist on the planet) skyrocketed in cost because of this grain price increase. Second, both the cost of running large farm equipment and the runaway price of natural gas–based artificial fertilizer synergized with the coincidental biofuels start-ups, which intensified existing regional famines in various parts of the world. According to UN records, global food prices rose 35 percent in 2008, escalating a trend that began in 2002.[20] Since then, prices for all food have risen an additional 65 percent. Last year, according to the UN Food and Agriculture Organization's world food index, dairy prices rose nearly 80 percent and grain 42 percent.

The intersection of more people and less oil is a recipe for human hardship. And if you think things are already bad, remove a significant portion of arable land—all of it swallowed by the sea. The result is an estimated four billion deaths between now and 2100, according to Peter Goodchild, creator of a popular Web site about this topic as well as books on sustainability, food production, and the way indigenous culture maintained food supply in the past.[21] Goodchild's is the most extreme human mortality estimate I have ever seen for this interval of time; he predicts that almost half of all people alive in this century will die. He made his tragic calculation without even taking into account either the effects of global warming on harvest yields or, more important, the reduction in farm area from sea level rise. Even a recent study (unconnected to the Goodchild work) on the effect of higher carbon dioxide on farm yields by David Battisti of the University of Washington did not factor sea level rise into his forecast of disaster in the near future. This reality must be factored in to any such calculations, and they have been utterly ignored. This is just another signal that rising sea level is just not on the scientific or political radar.

A CAR IN EVERY GARAGE

The millions of military conscripts coming home to America after World War Two returned to a country whose economic status and potential were not so different from those of China and India today. Consider that great emblem of Western wealth and emitter of carbon dioxide, the automobile. In 1945, far less than half of all American families owned an automobile.[22] Such is the case in India and China today. Both of the Asian powerhouses are set to explode economically, which will lead to a huge escalation in the ownership of private cars. To a lesser extent the same phenomenon is true of other soon-to-burgeon nations with large populations, such as Brazil, Indonesia, Russia, Thailand, Mexico, and Iran. Automobile manufacturers certainly know this, and they are planning to meet the new demand. What is unclear is the type of automobiles they will deliver—and what their average emissions will be. The answer to this economic question will in no small way dictate the "when" and "how much" of the rise in sea level we will experience.

Car emissions account for one-tenth of greenhouse gases worldwide. That fraction could grow greatly.[23] Globally, as of 2007, generating power produces nearly 10 billion tons of CO_2 per year, a quarter of all emissions, with the United States (home to more than 8,000 power plants out of more than 50,000 worldwide) responsible for about 25 percent of that amount, or 2.8 billion tons. Deforestation creates another billion tons, as of 2005. In 2004, personal vehicles created 618 million tons of carbon dioxide.[24] As citizens of developing countries get the itch for their own personal cars, that CO_2 count will certainly increase. The dramatic rise in car ownership will most likely ensure that emission reductions in other areas will be overwhelmed by many new tons of carbon entering the atmosphere. The result will be faster melting of the ice sheets.

Right now, on basis Americans annually produce 9 tons per person of carbon,[25] primarily through driving, and secondarily from warming or cooling themselves. Australians are worse, producing more than 11 tons of CO_2 emissions per person every year. Populous developing nations have far lower per capita emissions. For example, the average Chinese citizen produces 2 tons of CO_2 and Indians create about 0.5 ton per person. But if they all start driving cars, that will change. There are 9 personal vehicles per 1,000 eligible drivers in China and 11 for every 1,000 Indians, compared with 1,148 for every 1,000 Americans.[26]

U.S. automobiles and light trucks are responsible for nearly half of all greenhouse gases emitted by automobiles globally, according to "Global Warming on the Road." This new study by the Environmental Defense Fund also found that in 2004, the latest year for which statistics were available, the Big Three automakers—General Motors, Ford, and what was then called DaimlerChrysler—accounted for nearly three-quarters of the carbon dioxide released by cars and pickup trucks on U.S. roads. Carbon dioxide emissions from personal vehicles in the United States equaled 314 million metric tons in 2004. According to the report, that much carbon could fill a coal train 55,000 miles long—long enough to circle the earth twice. The same study reported that U.S. cars and light trucks were driven 2.6 trillion miles in 2004, the equivalent of 10 million trips from the earth to the moon.

How many new cars and light trucks can we add to this total before the earth gasps? In 2007, 56 million cars were sold in the United States, of which 15 million were small cars, with a total of all cars and trucks at about 800 million vehicles globally. In 2008, the Boston Consulting Group, a business think tank, published a report forecasting an additional 88 million cars and trucks worldwide in 2020—with 44 million of these coming from India, China, Brazil, and Russia alone. The Boston Consulting Group concluded that there will be two billion cars and trucks on the road at that time globally.

The future level of atmospheric emissions depends on what kind of cars we have by 2020. There has been much brave talk in America about the imminent change in Detroit from gas guzzlers to the "car of the future"—with most of the hype going to hydrogen fuel cell and plug-in hybrids.

It is way too soon to know whether the Great Recession of 2008–2009 will produce real change in the American auto industry toward more fuel-efficient cars. Yet, Americans may not be the most important players in this. As their countries industrialize and become more prosperous, the people of India and China will want Western luxuries—including personal automobiles. In no small way their success (or lack thereof) will dictate the fate of civilization—and the rate of sea level rise.

However, the current situation in India suggests that although the country's new, planned car models may not be large, they certainly will be more polluting than cars elsewhere in the world. My fear is that India, China, Indonesia, Brazil, and other countries where a vast new middle class is coming into existence—and thus is beginning to buy cars—will go the route of more pollution, simply because a "green" car with low or no emissions costs so much more than gas guzzler. With a median age of just under twenty-five and a rapidly expanding middle class, India overtook China in 2009 as the fastest-growing car market, according to estimates by CSM Worldwide, an auto industry forecasting service. India's new Nano automobile, which sells for only $2,500 (still a lot of money for the average Indian, but becoming more affordable each day), is thought to be the engine of the country's emerging auto industry, the car that will allow the average Indian family to have its own vehicle for the first time.[27] But to keep the car so inexpensive, its maker, Tata Motors, had to cut many corners, and reduced emissions was one of them. If it were an automobile driving American roads, it would actually be considered a very "green" car, because it is said to get 47 miles to the gallon on the freeway, which puts it right up there with the Honda Civic hybrid, the car I personally own, and the famous Toyota Prius. But the Nano's 47 miles add a great deal more CO_2 to the atmosphere than those of its Japanese cousins. Despite the Nano's relatively high gas mileage, environmentalists view the car as a disaster. They worry that a super-cheap auto will encourage Indians to act like Americans and rely on the car rather than mass transit. More drivers will add to air pollution, already a critical problem in more than half of India's cities, and compound the levels of carbon in the atmosphere.

The Chinese are behind the Indians in populating their roads with affordable private vehicles. But the Chinese also have a different economic goal than the Indians do. China wants to swamp the rest of the world with luxury-styled cars costing only $10,000—about a third of the cost of equivalent models being produced in Japan and the United States. And China aims only to meet U.S. emission standards—among the most lax in the world.

The net effect of the rise of the Indian and Chinese auto industries means more cars on the road by 2020 and beyond. Both countries' manufacturing bases have plenty of time to develop before the world reaches its population zenith of nine billion people between 2040 and 2050—and their auto industries will be able to deliver to a hefty percentage of those people their own shiny new polluting car.

To make those cars, both nations will need much more power than they currently employ—power that will come mostly from coal. Various think tanks now estimate that carbon emissions by China alone[28] will reach about 2 billion metric tons in 2020, an amount that by 2030 will rise to somewhere between 3 billion and 4 billion metric tons. (A metric ton is larger than a U.S. ton.) Alone among those of heavily industrialized countries, China's economy did not melt down in the worldwide recession in the latter part of the twenty-first century's first decade. In comparison, in 2007 the United States, the largest emitter of carbon into the atmosphere, produced 1.8 billion tons. Thus, China alone will more than double the United States' level of emissions in slightly more than twenty years. About 75 percent of this amount comes from burning coal, and the new power plants coming on line—by some estimates about three a week in China—have lifetimes of forty to seventy-five years, ensuring that enormous volumes of carbon will be pumped into the atmosphere until at least 2100, and perhaps beyond if we fail to develop cheaper and safer energy sources.

MORE PEOPLE, LESS ICE

The numbers reported in this chapter are stark, as is their message. We are producing too many people if we want to preserve the world we know—one that features ice sheets and the sea level around which we have developed a complex civilization. The rise in our numbers, more than any other reason, will ensure the rise of the sea—which not only will happen but also might be vastly underestimated unless we take immediate action. There will soon come a time when the ice reserves begin melting. All that water has to go somewhere. And that "where" is the heart of human civilization.

CHAPTER 4

FEEDING HUMANITY AMID
RISING SEA LEVEL

Northern Sacramento Valley, 2135 CE. Carbon dioxide at 800 ppm.

The horsemen rode south along the bed of McCarty Creek, their shot-guns loosely held but loaded and ready as they searched the long barbed-wire fence for the break that had to be there. It was grueling work under the intense Northern California sun. Off in the distance they could just make out Sutter Buttes, and there was the peak of Mount Shasta to the northeast, both stony excrescences bereft of snow. This small extinct creek, once fed by the nearby Coast Ranges, had eventually meandered into the mighty Sacramento River—when it had water, any-way. The Sacramento, that once-giant river, was itself just a trickle now, but come the autumn and winter rains it would again rise over its banks in winter flooding. No longer did any snow fall on the Sierra Nevada, a vast brown wall to the east. Where once there was snow until late into the summer, now it looked like a scene out of twentieth-century Nevada, where the basin and range topography was made up of brown mountains encasing valley fills of scree and cobble. The Coast Ranges and the Sier-ras commanded the vast valley that they rode through, one of the biggest valleys on Earth, in fact—the Great Valley of California. Beyond its size, however, its greatness was debatable.

The Valley, as it was still called, had once been one of the richest agricultural areas on the planet. It had been divided roughly in half by the Sacramento River Delta and the low marshes west of Sacramento. Its northern half had been farmed for fruit, olives, nuts, cotton, and especially rice, while the southern valley was once the largest vegetable-producing area on the planet. Now the Great Valley was bisected by the long extension of San Francisco Bay, which stretched all the way to Sacramento. Salt water from that enormous extension of the sea had gradually worked its way into the many aquifers that had once been necessary for irrigation, and every year the sea encroached both north and south into the major rivers of the Valley. Now, despite the intense engineering efforts Californians had put forth, most of those aquifers contained salt. But even that would not have been so bad had the climate continued to allow snow to fall prodigiously on the Sierras. Because the precipitation now came entirely as rain, there was no snowpack to melt and provide spring runoff just in time for sowing and watering new crops, or give budding trees a good drink in the first spell of hot weather.

That heat used to arrive in April, but now there was no winter here at all. In one respect it was a blessing—no longer did the characteristic and deadly early-morning fogs cause numerous fatal accidents on Interstate 5, the major north-south freeway through California, as drivers rear-ended others in the pea soup. There was no fog at all now, because the tropical temperatures of the Valley never rose to the dew point. But the lack of fog was of little importance to drivers, because there were none on the freeway except for truckers. Personal automobiles had been outlawed some decades before, in a vain effort to save some of the world's oil. Yet goods still needed to be moved from place to place, and people needed to travel as well, thus swelling the freeways with buses and trucks.

Long before the Great Valley had gone bone dry in the new climate, thanks also to a nearly water-free summer, its farmers had switched from planted crops to four-legged ones. The winter rains and strong sun still allowed massive quantities of grass to grow, so giant cattle and sheep farms had flourished where rice, vegetables, and fruit trees had once thrived. But soon the climate became too dry for even the sheep, the grass too short and too short-lived. Despite the ranchers' efforts, no cattle or sheep could survive the brutal summers.

America still needed meat. Even if it could not grow cows or even sheep, the Valley could sustain other quadrupeds that were well adapted to deal with drought, and still give meat. First the ranchers had tried white-tailed deer, then antelopes. But for speed of growth and ability to withstand dryness and heat, kangaroo just could not be beat. And out of that recognition had grown the last large source of ready meat in markets across the land. As always when the horsemen ventured out, vast flocks of crows circled overhead, big bastards. Gone were the old Valley buzzards, wiped out by the crows. In fact, all the Valley birds were gone or nearly so, partly because of the loss of the once-verdant vegetation and fields, but increasingly because eggs and newly hatched songbirds were consumed by the ever-increasing number of crows. The crows, the smartest of all birds, were quickly evolving into tight social structures, and now only one male would breed with the females of the flock; the large alpha male led all. They had found that bird eggs made great meals, and when the birds were gone they took to attacking larger game. Already there were three or four known cases of small children being pecked to death and partly eaten. The war between humans and crows was now joined in earnest, and it was partly for these that the horsemen kept their shotguns loaded, with birdshot.

But also the men were wary of the packs of wild dogs, strange canines that had mingled with coyotes, with not a few coyote-poos, coyote shepherds, and pit coyotes, all looking ridiculous, but still deadly in a pack. No one could keep carnivorous pets anymore, with meat so scarce and the carbon footprint of even a small dog greater than that of owning and operating an SUV, but many city dwellers raised dogs for food, a new practice learned from the immigrant Polynesians and Vietnamese.

Yet if the men remained armed for varmints, another unsaid reason was that they needed defense—and sometimes offense—against a two-legged threat. Vast bands of dispossessed humans from the Southern California megacity of Los Angeles–San Diego continually migrated up through the Valley toward the wet, productive lands of Canada, where all assumed that work could still be had within one of the foremost economies on Earth. Canada and Russia were now the richest countries on Earth, and they guarded their borders ruthlessly. But that did not deter the truly desperate.

America was hungry. But not as hungry as some other places, namely China, Northern Africa, and India. China was the home to major famine, because the annual snowpack in the Himalayas was also affected by the warming. And many of the major Himalaya-bred rivers in China and India, including the Indus, Ganges, Brahmaputra, and Yangtze, had radically diminished their flow because of both the change of snow to winter rain and the final melting of every one of the mountain glaciers that had also been an important source of seasonal water and input into the major rivers.

But perhaps even more disastrous for the world was the loss of all its deltas, like the Sacramento River Delta here. These low, swampy places had never seemed to be the best real estate on the planet. In places such as the lower Mississippi Delta, the Mekong, and the Indus and Brahmaputra, the intersection of a major river with the sea was manifested as swampy, shifting land. The formation and abandonment of river channels in these areas—occurring in as little as a year or two—made development difficult, if possible at all. But in one regard the deltas were priceless. They had been made up of the best soil on the planet. Now they had disappeared into the sea.

It was not as if China, India, and the rest of the world had not seen the impending disaster of water—and hence of food. Way back in 2007, the formerly giant Yangtze River fell to its lowest level since record-keeping began, in the 1860s, as the most serious drought of two centuries maintained its decade-long grip (and counting) early in the new century on the entire Yangtze Basin and its nearly 1 billion residents, surpassing in severity even the deadly coeval droughts in South Africa, Australia, Argentina, and the entire Midwestern and Southwestern United States into Mexico.[1] The Chinese had tried to compensate by discharging water from the relatively new Three Gorges Dam, but that had accomplished little.[2] Decade by decade, that enormous feat of human engineering had become an enormous feat of human waste and stupidity, as the dam rapidly silted up.

It was as if the gods were just plain nasty, seeming especially unkind toward China. In those places needing more water, river flows diminished, while those in wet regions awash in water increased too much. In some northern arid provinces, such as Ningxia Hui Autonomous Region and Gansu Province, the situation was already dire in the first decades of the twenty-first century and became more so, while the too-wet southern provinces, such as Hubei and Hunan, suffered ever-larger floods and flood-induced mortality, either from drowning or from famine caused by diminished farm yield resulting from loss of topsoil. In northern China, especially in Ningxia and Gansu, water resource per capita had decreased but remained sustainable until 2050. After that, however, disaster after disaster had hit as a succession of heat waves and droughts killed millions. But even the farmers in Ningxia and Gansu were relatively well off compared to the people of Inner Mongolia, which had become a Sahara uninhabitable to humans—just as the entire equatorial region of the world had become steamy centers of human misery and squalor under temperatures 10 degrees hotter than the already formidable heat of the late twentieth and early twenty-first centuries. The modern tropical states such as Singapore and Indonesia had reverted to their roots because air conditioning had become too expensive to support.

China had avoided complete disaster for a while by calling due the many and enormous loans it had made to America. The Americans were no longer consumers of Chinese knickknacks, clothing, and toys, and when China demanded payment, the country was forced to pay in grain, at a huge cost to the U.S. standard of living, which by 2100 was just a fraction of what it had been a century before. Eventually the Americans had just quit paying, their nuclear weapons still formidable and many of them aimed at Beijing. Thus, the Americans counted themselves lucky; at least they were fed. In China, over 100 million people starved to death in one particularly harsh decade of drought and famine brought about by the lack of river water—and by the loss of coastal rice paddies that were victims of the rise in sea levels—which were now a devastating 3 to 5 feet above their levels in 2000.

SEVEN BILLION MOUTHS TO FEED

Although our species has been present on the planet for 200,000 years, we did not really begin to rise in number until 10,000 years ago with the start of agriculture.[3] The simple act of putting a seed into a prepared field and nursing it through germination and pollination to maturity transformed human life. Agriculture brought about the first cities, accompanied by the organization of large populations—including the first armies. Farming also brought global epidemics, the effects of large numbers of people for the first time living in such proximity—and fouling their nests in the process with human and other wastes, as well as enticing such disease vectors as rodents and fleas.

We have come a long way in 10,000 years. Anyone even in 1900 would disbelieve that our world not only contains almost seven billion people, but that we are successfully feeding a large majority of that enormous population. Yet feed the world we do (mostly, and sometimes not completely at each meal). How is this feat accomplished? The miracles of large-scale farming have contributed, as has the biological breeding of hardy and high-yield crops and livestock. The use of fertilizers is a factor, along with the invention of the internal combustion engine and its use in agricultural machines. Finally, the highly successful technology preserving food for long periods and the infrastructure to move it large distances in short periods of time are also part of the equation.

How much food do we make? One of the best estimates is from the U.S. Department of Agriculture, which says that in 2008 and 2009, the world produced 600 million tons of wheat, 1,100 million metric tons of coarse grains, and 430 million tons of rice, a sum of over 2,000 million metric tons of grains. The area needed to produce the wheat was 217 million hectares of land, 316 million hectares for coarse grain, and 157 million hectares for rice.[4] Yet while impressive, do these figures add up to the needs of an ever more populous world? Ominously, none of the estimates about future food needs have factored in the additional pressure on agricultural production from the growing demand for biofuels as well as the loss of prime agricultural land from sea level rise. The climate extremes that are coming with warming—droughts and floods—will also exacerbate the situation as we strive to keep feeding ourselves.

Over the past few years there has been a worldwide reduction in grain harvests, probably due to environmental degradation from decades of unsustainable practices of the so-called Green Revolution, where various countries tried to increase productivity through seemingly easy fixes—using genetically engineered crops rather than doing the harder work of initiating sustainable agriculture suited to the particular land being farmed. The result has been massive soil erosion, loss of soil fertility, loss of agricultural land through salinization, depletion of water tables, and increased pest resistance.[5] Other environmental costs of the Green Revolution include surface and groundwater contamination, release of greenhouse gases (especially through deforestation and conversion into agricultural land), and loss of biodiversity.

PLANTS IN A HIGHER-CARBON ATMOSPHERE

Anyone who has tried to grow plants either indoors or outside knows how tricky some can be, how resistant others are to the grower's ministrations. Too much or too little water, too much heat or cold, the wrong kind or amount of nutrients—all these climatic factors can spell quick or drawn-out death to any given plant. And not just the absolutes kill; for many plants the simple process of change, especially rapid change, is sufficient. With the prediction that the earth will warm some significant amount, and so alter patterns, amount, and types of precipitation, botanists are scrambling to figure out which plants might suffer most, and which others might thrive.

On its face, you could assume that for plants, global warming would be a win-win situation. As long as some upper thermal limit is not reached, slightly warmer temperatures actually stimulate faster cell growth in plants, and the increase of carbon dioxide in the atmosphere should also be beneficial, because CO_2 is their source of carbon, and thus the building block for all plants. In fact, on a warmed Earth there will be winners. Early research suggests that the increased CO_2 will enhance the yields of the most important plants of all, at least to us humans: the crops that sustain us. According to this point of view, a doubling of atmospheric CO_2 could lead to crop increases (all else being equal) of nearly 40 percent.[6]

But one group of American scientists has been watching with growing alarm a second series of experiments exploring the prospects of agricultural yields in the face of rising temperatures and increased CO_2. In a new study, David Battisti and a colleague at the University of Washington found that high temperatures do cause such plants as rice, corn, and wheat to grow faster but also reduce plant fertility and grain production.[7] With average growing-season temperatures expected to rise more than 6 degrees Fahrenheit in many places, crop yields will fall 20 percent to 40 percent, the report estimates. The effects will be aggravated by increased evaporation and loss of soil moisture. Even in the United States, where warming caused by greenhouse gas emissions is projected to increase some crop yields through the middle of this century, harvests will most likely fall by 2100 as the heat intensifies. Worldwide, Battisti points out, subsistence farmers and the poor will feel the impacts most keenly.

The Battisti report is a wake-up call for us to develop new heat-resistant crop strains. Yet it can take two decades or more to breed a new strain, and for the past several decades, investments in agricultural research have stagnated. Although new kinds of philanthropic organizations, such as the Bill & Melinda Gates Foundation, are helping fund an effort in Africa to develop hardier crop strains, it turns out that this work hasn't focused specifically on heat tolerance.

These newly presented factors—lower yield in a more populated world, coupled with the estimates that various midlatitude food producing areas, such as the American Midwest, will experience more and larger floods—will have a profound impact on agriculture worldwide. One major case in point comes from the West Coast of North America, where the current loss of snowpack will lead to major drought each summer, and major flooding during the winter, wreaking havoc on one of America's primary agricultural areas.

As important as they are, however, floods are just one of the unquestioned effects that increasing global temperatures will bring about. Another major consequence is biological change caused by warming and sea level rise. Because temperature is one of the major factors affecting chemical reactions, and because any kind of life is nothing more than a factory of multiple, continuing chemical reactions, it is quite logical that myriad changes would happen. But the impact will go far beyond individual physiological pathways. Because organisms organize at higher levels than at the individual level, including at the species level as well as into various kinds of ecological units, the effects of global warming will be biologically omnipresent. It is no surprise that we can expect changes in agriculture commensurate with the scale of temperature change.

That being said, I was somewhat surprised to learn, while researching this book, that there has been relatively little research into the actual effects of a changing climate on various crops. There has been particularly minimal inquiry into how individual countries will fare in terms of crop yield. But William R. Cline presents a sobering analysis of the "winners" and potential losers in the vital ability of individual countries to feed themselves and, if they are lucky, to export any surpluses in the face of climate change and (for those bordering oceans) sea level rise.[8] Cline is no newcomer to such analysis: his pioneering 1992 book *The Economics of Global Warming* set the table regarding the economic effects of climate change.[9] His new analysis is welcome and timely, and much of the following discussion is based on his findings.

As we've seen, if there was ever any good news about global warming and rising carbon dioxide, it was always the possibility that future rises in both would increase crop yields. The vision of currently cold regions producing, either in larger amounts or even for the first time, bountiful harvests of grain, corn, vegetables, potatoes, rice, and edible livestock was hopeful to many analysts. But sobering research that indicates that many crop species would actually produce less under higher temperatures and carbon dioxide loads seems to balance the unquestioned positive effects of having new lands become agriculturally useful. Also tempering hope is that the warming is a double-edged sword—though it may create new regions of farmland, it could make currently productive cropland useless to agriculture. The net result, based on current models, is that overall effects on agriculture will be negative by 2100—and then become ever more so.

What will these effects be, and where will they most dangerously occur? The consensus is that negative aspects will be greater in developing countries than in those already designated as developed, where one aspect is a modern and effective agricultural system. Far more countries at low latitudes already are warm relative to the rest of the world, and thus they are closer to thresholds where further warming will cause reduced yields of current crops. Second, agriculture characteristically makes up a larger proportion of gross national product in these countries, which means that losses in this sector will have a disproportionate effect on national economies, with higher possibilities of political perturbations resulting.[10]

Unfortunately, the nature and timing of global warming does not enhance planning for, or mitigation of, any predicted global change. Part of the reason for this is physics: today's emissions will not play into the warming of the ocean for a few decades.[11] Also, planning for effects that will kick in many decades in the future is not the sort of thing that governments do. Day-to-day problems generally tax all public planning resources, or governments plan ahead at most for several years, not decades hence.

Nor are the impacts themselves necessarily obvious. While future consequences of global warming on irrigation and international agricultural trade are clear enough, the important issue known as "carbon fertilization" is relatively unknown to the populace at large. Carbon fertilization refers to the most basic aspect of plant physiology—the means through which a plant obtains carbon, the basic building block of cell architecture. At this level whatever benefits come from rising carbon dioxide occur. With more CO_2 molecules in the atmosphere directly surrounding any plant, it can more easily acquire carbon, essential not only for the plant's structure but for energy acquisition as well. In all plants, carbon dioxide enters the plant through small holes in leaves and sometimes stem material called stomata. The greater the carbon dioxide, the fewer the stomata.[12] This enduring ratio has allowed scientists to estimate the CO_2 in ancient atmospheres through the analysis of fossil leaves, because even ones that have been modernly preserved show the relative number of stomata. Comparing this number over time allows data on the relative direction of CO_2 in the atmosphere, as well as measurements relative to today's plants. And yet necessary as they are, stomata can also adversely affect plants, because just as gas goes in the plant, so too can its liquid water pass back out. Too much water loss is fatal. There is more water loss under conditions of low CO_2, and therefore more stomata, than when the atmosphere's CO_2 is high and the same plant can get by with fewer stomata and better water retention.

There is a marked difference in the effect of higher CO_2 among various crops. Botanists break down plants into two types.[13] The first uses a so-called C_3 metabolic pathway during photosynthesis, in which one of the crucial chemical reactions leading to energy acquisition employs three separate reactions of a newly acquired CO_2 molecule and the energy acquisition centers in the plant. The second, the C_4 plant, creates the same reaction using four such cycles. From a geological perspective, C_4 plants are far younger, appearing on the scene in large numbers only during the Paleocene period of 65 million to about 20 million years ago. The reason for plants' evolution into the C_4 type is a bit ironic given what is happening in our atmosphere. As we saw in Chapter 2, the long-term trend in atmospheric carbon dioxide levels has been downward, and the C_4 plants fare better in a low-CO_2 environment than do C_3 plants.[14] In modern agriculture, C_4 plants include maize (corn varieties), sorghum, millet, and sugarcane, as well as the many grass varieties fed upon by grazing and economically important mammals such as cows and sheep. The C_3 plants of agricultural importance include the staples of human food such as rice, wheat, soybeans, fine grains, legumes (vegetables), and most trees including all fruit trees.[15]

Studies on various species of both C_3 and C_4 plants have been conducted on small scales, using closed chambers in which the various gases and light, as well as moisture and temperature, can be manipulated.[16] It was through such studies that initial findings of higher growth rate, and ultimately yield, for many of the plants listed above gave rise to a muted but real sense of hope that higher CO_2 might encourage greater plant growth. In studies conducted more than a decade ago, CO_2 levels of about 550 ppm (compared to the current atmospheric level of 385 ppm) resulted in a yield increase on average of 11 percent for C_3 crops and 7 percent for all crops, regardless of whether they were C_3 or C_4 plants. More recent studies, however, dispute these early findings. In fact, no increased yield was noted in any of the C_4 plants, while some of the important C_3 plants, most notably wheat, showed higher yields than in the older studies. Since C_3 crops account for three-quarters of global agricultural yield today, this seemed like a hopeful tradeoff. Estimates of when atmospheric emissions would hit the level of 550 ppm range from the mid twenty-first century to a decade or two thereafter.

Unfortunately, when CO_2 values were raised to levels that might be present by 2100—that is, 750 ppm (which will be the case unless a major reduction in carbon emissions is instituted globally)—the bottom line is not so favorable. There are still increases in crop yield—averaging 15 percent—but these would very likely be more than offset by alterations in global temperature, precipitation, and perhaps most important (yet rarely accounted for in the models), the loss of land area to both excess heat and to inundation from the sea, or to salinization of the soil (or both).

It is not just this change in carbon fertilizing that must be accounted for, of course. As noted above, temperature rise will affect both irrigation and, ultimately, trade. Irrigation might be the most critical aspect of food production in a warmed world. As temperature rises, plants tend to lose more water, and thus benefit from the greater water in the atmosphere of coastal (lakes and the sea) farms brought about by global warming. But in those locales far from larger bodies of water, the higher temperature increases the dryness of air, and this negatively affects plants. With drier air, crops require more water than they would in more moist air, not just because the plants are losing more water than at lower temperatures, but because hotter, drier air causes the soil to lose ever more irrigation water, so more water is needed for the same acreage. As we will see, as sea level rises, access to enough freshwater for both human consumption and crop irrigation will be a major challenge in the future.

In some cases, scientists predict that higher temperatures will be accompanied by higher precipitation, and thus no net change will be necessary in irrigation volumes. But this is the minority situation. If we jump to the year 2080, a time when annual U.S. farmland temperatures will have jumped by a mean of about 8 degrees Fahrenheit, the prospect is quite different. Climate models indicate that this substantial increase in heat will be accompanied by a slight reduction in mean annual rainfall (only 4 mm, or about a quarter of an inch) for the United States as a whole.[17] Nevertheless, the substantial increase in temperature will have a huge effect on irrigation needs: the amount of water needed to irrigate the same area of cropland being farmed today would be 20 percent higher than current volumes. At the same time, the amount of water available in the United States is projected to be reduced substantially, especially in those areas dependent on annual glacial melt and snow thaw discharged into rivers. By then, glacial pullback will have substantially reduced the size of rivers in many areas, especially those fed by glaciers in the Sierra Nevada.

The loss of irrigation water due to higher temperature has rarely been figured into the cost of agriculture in the future. But any change in irrigation is costly, because of the substantial costs of irrigation infrastructure, from retrofitting to the building of new pipelines, aqueducts, and the like. Moreover, the increase in temperature will also cause some areas that currently require little or no irrigation to change to hotter, drier conditions demanding major irrigation efforts—and brand-new infrastructure to move water.

While this particular study related to the United States, its implications affect the rest of the world as well. For America and all other nations, two fundamental questions remain unanswered: who will pay for the new irrigation infrastructure, and where will the extra water come from?

Finally, we must address the question of whether international trade can somehow make up for any potential regional or national losses in food production as a result of climate change. Currently an almost smug conviction prevails that trade will more than mitigate food losses in various parts of the globe. Much of this optimism comes from the developed nations with current food surpluses, whose leaders seem to believe that less developed countries—those the United Nations considers to be most at risk for reduced harvests—can simply buy more food from them. In fact, the economics suggest a far different outcome. First, because agriculture is a larger proportion of GNP in less developed countries, losses in crop yield will reduce their ability to buy their way out of shortages.[18] Second, a reduction in worldwide agricultural yield shifts the supply down without an equivalent loss of demand. All this spells trouble. Such data show that trade cannot be part of the equation to solve the crop shortfalls that probably will occur by 2050 and certainly by 2100.[19]

OTHER MODELS FOR CROP YIELDS

How much will the changing climates in various countries affect their food production? The Intergovernmental Panel on Climate Change (IPCC) has come up with various climate scenarios—models based on Global Circulation Models (GCM), discussed in Chapter 1, that give us an accurate and alarming forecast of who will eat and who will not. Because the IPCC is notoriously conservative in its predictions, the following results have to be considered as minimum predictions.

Basically the IPCC models analyze alterations in temperature and precipitation in a given area, called a cell. The current model resolution has the surface of the earth divided into 259,200 cells, which means relatively small areas can be modeled. These results are then coupled with known environmental effects that various crops are prone to, and thus predict future yields by country. Clearly this kind of modeling works best for smaller, agriculturally homogenous countries, whereas results for agriculturally diverse giants such as Russia, Canada, the United States, and Brazil require a number of estimates. Because of the heterogeneity of local climates across these countries, they and other large nations have been divided into subzones, each of which has its own estimates. Significantly, the IPCC models project that mean annual temperatures exceeding about 54 degrees to 57 degrees Fahrenheit will force a loss of crop yield. This estimate comes from a study in 2000 by a research team led by Robert Mendelsohn and is based on the physiological effects of higher upward temperatures on crop-yielding species.[20]

The IPCC models take mean temperatures by country from 1960 to 1990 and compare them with temperature changes predicted for 2070 to 2099. The results fully corroborate the prediction that rising temperatures will have a far stronger impact on developing countries than on developed ones because of the latitudinal distribution of the two groups' populations. The results showed that sixty-two developing countries (or subzones within them) exceed the 54–57 degree threshold, while twenty-five do not. In contrast, only seven developed countries or subzones within them exceed the temperature threshold, while twenty-two are below it. The ultimate result is that the rich countries get fatter and the poor ones go hungry.[21]

The IPCC also modeled a predicted increase in precipitation.[22] As we have seen, the one bright spot in the intersection of rising temperatures and plant physiology is that increased precipitation will offset the negative effects of high temperature. Unfortunately, the 2007 IPCC reported that for the world's Global Circulation Models, most tropical areas have increased mean precipitation, while most of the subtropical areas have decreased mean precipitation. In high-latitude areas, on the other hand, precipitation increases.[23]

Unfortunately, it is precisely the subtropics—the temperate regions—that today feed most of the globe, and it is in these vast and currently rich agricultural sites that rainfall will decline while temperatures increase. This will necessitate considerably more water for irrigation, even as the overall volume of available water is significantly diminished.

Robert Mendelsohn and his colleagues conducted a study in 1999 that looked at land yield in terms of the money each acre could produce and computed that number on a graph of temperature plotted against precipitation.[24] As noted above, ever-higher temperatures require higher precipitation to maintain crop yields. Areas with very low precipitation can grow crops only at temperatures between 45 and 57 degrees Fahrenheit, whereas higher precipitation allows farming in areas both hotter and colder. At very high temperatures, above 82 degrees, for instance, crops require a mean precipitation of at least 12 mm per day, or half an inch of rain each day, year round. Very few places are that wet.

Using this model, Mendelsohn and his colleagues went on to compare predicted crop output for various countries, including the United States (which because of its size was subdivided into six regions, each with its own predictions), for the year 2080. The results for the United States are startling. Three areas, the Northeast lakes district, the Rocky Mountains and contiguous Great Plains, and the Pacific Northwest, all show gains in crop yield. But the Southeast, the South Pacific Coast, and the southwest plains show such significant losses that the United States as a whole can be predicated to have reduced crop yields of just under 15 percent of today's output. Coupled with the dramatic increase in America's population expected by 2080, this clearly is not good news.

Even worse news is that all of these various models and predictions assume that current crop areas will remain above water and free of plant-killing levels of salt in the soil and groundwater. By 2080, the expected minimum 3 feet of sea level rise will greatly affect the Pacific Northwest, Southeast, and South Pacific Coast agricultural regions. Thus the 15 percent loss of yield is a very conservative estimate; ultimately it could be far greater. There is a very real possibility that by the end of this century, the United States will not be able to feed its people without importing food.

TABLE 4.1. Predicted percentage crop yield gain or loss by 2080[25]

Area	Percent change crop yield
Africa	-15–20
Argentina	-4
Australia	+3
Brazil	-16%
Canada	+12.5
China	+3
Europe	+4.5
Far East Asia (excluding China)	-15
Indonesia	-5
Latin America excluding Mexico	-19
Mexico	-24
Pakistan	-26
Turkey	-12

CROP FORECASTS FOR DEVELOPING COUNTRIES

As bad as things might be in the United States by the end of the twenty-first century, its predicted crop yields by 2080 are still much more favorable than those for many developing countries. India faces a dire future, according to IPCC models for its four subregions.[26] Net output of crops in monetary value will drop 31 percent for India as a whole but a whopping 61 percent for the northeast region of India, where climate is already not so kind to agriculture. Once again, these figures do not factor in the loss of coastal agricultural areas to sea level rise.

Another set of projections, the Rosenwig/Igelsius model, is more conservative but still is extremely troubling.[27] Regional and selected country results are discussed in the primary article.[28]

WHO EATS, WHO STARVES

In Chapter 3 we looked at the influence of world oil resources on agricultural yield. Let us revisit this subject, but at a time in the future when sea level has risen significantly.

It is doubtful that the famines predicted to accompany sea level rise will be worldwide. There will always be richer and poorer countries. But how will any possible famines be distributed? One way of predicting at least the locations of future famines is to examine current "famine hot spots" around the globe. We do not have to look very far beyond Africa, because that continent has been, and is predicted to remain, the ground zero of famine.

Prolonged civil wars in the Democratic Republic of the Congo (DRC) and in Africa's Afar, along with civil unrest in Kenya, have led to de-creased agricultural yield, while the unrelenting tyranny of Zimbabwe's dictator, Robert Mugabe, has destroyed his country's agricultural base. Reduced crop yields accompanied by a rising population mean that Tan-zania, Mozambique, and the DRC are likely to face serious shortages by 2030, according to a comprehensive 2008 study by a team from the Swiss Federal Institute of Aquatic Science and Technology.[29] This group as-sessed the impact of climate change by 2030 on the production of six ma-jor food crops in sub-Saharan Africa: cassava, maize, wheat, sorghum, rice, and millet. Two environmental aspects are deleterious—the pre-dicted higher temperatures and lower rainfall totals. Higher tempera-tures will make wheat wilt, with yields falling by up to 18 percent.

These assessments, when combined with projections for population and economic growth, led to a series of predictions about which African countries would be well fed, and which would not. Tanzania, Mozam-bique, and the DRC fared worst for food security. These countries will see the first major famines. But soon afterward, many more countries will not be able to feed themselves. That might mean a billion dead by 2200.

THE SACRAMENTO DELTA: A LARGE-SCALE EXAMPLE

The word "California" invokes many images, which befits a state so di-verse, and sunny beaches and long, picturesque coastlines are surely cen-tral to that vision. But anyone who has made the long drive from the Oregon–California border south on Interstate 5 to Los Angeles knows a different California—two different states, in fact. After a couple of hours of southbound travel, the beautiful mountain views on either side of the busy freeway give way to a horizontal landscape. The land settles and flattens and the mountains recede to the east as the rugged Sierras and to the west become the lower but still impressive coast ranges. But soon both sides all but vanish into the immense, flat valley, which seemingly grows ever larger, and the views soon resemble those found in Kansas, or Nebraska, or any other Great Plains state: endless grasslands punctuated by neat and orderly fields. Yet the mountains' presence is still felt. From both sides flow many small creeks that, during the brief rainy periods of winter and especially in the first heat of early Sierra spring, cause huge volumes of water to flow into the valley. The small streams join bigger streams to flow into the mighty Sacramento River. Over hours of driving, this view persists, a flatness with occasional views of the south-flowing river, the first three hours in what is called the Sacramento Valley, which becomes the San Joaquin Valley, the division caused by a wide expanse of river, running shallow over bank and levee deposits, and rice fields around Sacramento. Then, after this brief bout of interesting scenery, it becomes flat fields again, with the occasional view of giant aqueducts and irrigation systems as well as another central river, the San Joaquin, not as large as the Sacramento but still considerable, especially in the spring when the mount caps of Yosemite and the other parts of the southern Sierra shed their snowpack in the hot sun.

This vast expanse of land, loosely squeezed on its eastern and western flanks by uplifted terrain, is named the Central Valley of California, but to residents of the Golden State it has another name: the Great Valley. Once that name referred only to its size and its unimaginable length to those who had to journey from one end to the other in horse-drawn wagons, but later that name evoked the wealth of agriculture that it brings to California. In fact, the vast valley stretches some 400 miles, from its northern end near Redding to its final length just before the famous stretch of highway known as the Grapevine. Although dotted with towns, the valley is largely empty of people. But any long drive shows that very little of this valley is still cloaked in its original verdure of prairie grasses and stunted cottonwoods and gnarly oaks along the Sacramento River in the north half, and the many smaller tributaries flowing into the San Joaquin Delta in the south.

The modern history of what is known simply as the Delta reveals profound geologic and social changes that began with European settlement in the mid-nineteenth century. After 1800, the Delta evolved from a fish-ing, hunting, and foraging site for Native Americans (primarily Miwok and Wintun tribes), to a transportation network for explorers and settlers, to a major agrarian resource for California, and finally to the hub of the wa-ter supply system for San Joaquin Valley agriculture and Southern Cali-fornia's cities. Central to these transformations was the conversion of vast areas of tidal wetlands into islands of farmland surrounded by levees. Much like the history of the Florida Everglades, each transformation was made without the benefit of knowing future needs and uses.

As originally found by European explorers, nearly 60 percent of the Delta was submerged by daily tides, and spring tides could submerge it entirely. Large areas were also subject to seasonal river flooding. Once this expanse was too cold in winter and far too hot in summer for any human use except raising vast herds of cows and sheep. But the Roo-sevelt administration's New Deal put legions of Depression-era work-ers back to work on vast constructions, first taming the Sacramento River from its long southern meandering to its hard right turn toward the sea near Sacramento, with its final rush westward to the 80-mile-dis-tant San Francisco Bay and its salty ocean egress underneath the new Golden Gate Bridge. President Franklin D. Roosevelt authorized the Central Valley Project on December 2, 1935. Originally the project con-tained three divisions: Friant, Kennett (also known as Shasta), and Con-tra Costa. The Rivers and Harbors Act of 1937 reauthorized those divisions and prioritized improvement of navigation, regulation, and flood control of the Sacramento and San Joaquin rivers, with irrigation and domestic uses secondary. On July 1, 1937, the name of the Contra Costa Division was changed to the Delta Division and included the Contra Costa Canal and the San Joaquin Pumping System. Not content with the many new east–west diversions in the northern half of the Great Valley that resulted in copious water flow—allowing fields of wheat and, nearer Sacramento, the then-odd crop of rice—the engi-neers set their sights on bringing water to the far drier San Joaquin Val-ley through the construction of monstrous culverts. With water overpasses and underpasses, aqueducts, and creek courses of all kinds, they succeeded admirably. And so they changed the nature of agricul-ture and food production in America.[30]

As had happened earlier in the Sacramento Valley, nearly worthless flat grassland suddenly had water—and value. The giant cattle farms gave way to fruit and eventually vegetable crops worth far more. The sun still baked from April through October, but with the copious water now available crops quickly grew, were harvested, and were replanted. The rate of growth was phenomenal; rail lines carried the fruit and vegetables south to Los Angeles or north to Sacramento, where the produce threaded everywhere else across the country. No longer were fresh vegetables only a treat of fall; now they appeared in markets before the Eastern and Midwestern snows had gone, and at prices so low that everyone could afford to partake of them. It was a cornucopia beyond imagination, and soon the San Joaquin Valley was producing as much as a quarter of all fruit and vegetables for the entire North American continent.

From the 1930s onward, this bounty made the ever-fewer owners of the ever-larger farms ever richer. And Americans benefited from fresh beans, peas, corn, squash, tomatoes, cantaloupe, and on and on instead of the mushy canned goods that for several generations had been the main source of the vitamin-rich plant parts of the human diet.

Not only farms prospered but livestock as well. The new water let new and more efficient cattle and sheep ranches establish themselves in the areas too far from the main aqueducts to grow thirsty vegetables. Huge sprinkler systems caused grass to grow rapidly, allowing sheep and cattle to fatten quickly, and soon the giant stockyards and slaughterhouses centered in the long, sleepy San Joaquin Valley town of Coalinga provided new employment while creating a stench that spread up and down the entire San Joaquin. Truly a paradise came to the desert. But as in any paradise, serpents lurked in the garden. The biggest of these was a simple chemical compound with enormous ramifications in human history and greater implications for the human future: salt.

The thousand miles of levees built around the intersection of the Sacramento and San Joaquin rivers changed what had been a vast marsh into a gigantic man-made and human-controlled lake. The Army Corps of Engineers led the creation of dozens of delta islands as they drained the marsh. Now in the San Joaquin portion of the delta region, a system of designed channels and pumps carefully manages all the precious water. But the Delta lies at sea level, and even though it is far inland from the salty seawater of San Francisco Bay, the great tidal change found along the San Francisco coastline pushes seawater all the way to the Delta, where it mixes with the freshwater from the two rivers. Sherman Island, one of the largest islands in the Delta, sits at the confluence of the Sacramento River and the bay, where the salt water meets freshwater. Before the waters meet and become unusable for either humans or agriculture, vast pumps commandeer the river water and transport it either south, toward the 3 million acres of agriculture in the San Joaquin Valley, and 21 million people in Southern California, or west, to 4 million people in the Bay Area.[31]

There has always been a fragile balance between the salt water and the fresh, with us humans doing everything possible to keep as much salt out of the Delta areas as possible, for obvious reasons.[32] Even slightly salty water cannot be used. It requires constant management and enough freshwater at all times to push the salt water back. Any change in sea level will radically change that water chemistry and dynamic. Even the most conservative estimate of future sea level change—a rise of 1 foot by the year 2050 and 3 feet or more by the end of the century—will mean trouble for the levees, rocks, and dirt mounds that keep the water in its place.

According to Jeff Mount of the University of California at Davis, there are two types of levees: those that have failed, and those that will. While seemingly tongue-in-cheek, this statement probably contains a sad truth. On Sherman Island, wind-driven waves lap up against a rocky levee. During a typical storm, with extremely high tides, waves rise as close as 12 inches from the top of the levee. In an NPR interview, Mount noted: "It's a game of inches out here. You're just sort of clinging to the edge here, with very little margin for error. Regrettably the sea level is rising. So, that's going to go over the tops of the levees much more often in the future."[33] The Delta islands are not permanent structures; already they lose about an inch of elevation a year, as soil is oxidized and blown away. "We may be as much as 15 feet below sea level," Mount added. "And just on the other side of this levee is water that is at or above sea level, and it is trying real hard to get in here. And it is just that crummy little levee that is keeping it from getting in here." The water will come through in one of two ways, he said. "It's going to do it gradually—sea level rise and changes in inflows—or it's going to do it suddenly through the collapse of the levees." And if there's a major levee collapse, water will rush in so quickly that it will suck salty water out of the bay and into the Delta in what Mount called "the big gulp." He is in awe of the water's force: "Just the noise of the water rushing into this island, and it's the sound of like a waterfall as this rushes in, and scours this hole in the ground as the water rushes in, and hurling pieces of soil way out onto the island. I mean, the power of these levee breaks is immense, unimaginable, and there's nothing you can do about it."

SALT INTRUSION

The intersection of the ocean and its inflowing rivers, where salt water meets fresh, is always a dynamic region of high energy. The San Francisco Bay–Sacramento Delta system is a case in point. Left to its own devices, it would maintain equilibrium thanks to natural forces. But once the San Joaquin and Sacramento valleys became the home of multibillion-dollar agricultural industries, nature was dealt out of the game—or so we humans thought. But salt was an enemy that human intervention made even more insidious.

The rise and fall of sea level throughout the ice ages has caused many changes in the amount of salt making its way up the rivers and into the vast inland valleys of California. Ever since humans started coming to the region in droves, that flux has been a source of both irritation and economic distress to agriculturalists as well as townspeople. During the twentieth century, before the Army Corps of Engineers got to work, the salt water intruding from San Francisco Bay and Suisun Bay caused problems in the valley, especially for the towns of Antioch and Pittsburg. Unless freshwater from the rivers flowed past Antioch at a minimum rate of 3,300 cubic feet per second against the high tide from the bay, salt water would enter Suisun Bay and the Delta, lowering the water quality. Between 1919 and 1924, salt water in Suisun Bay caused wholesale biological changes, including invasions of the devastating, wood-boring invertebrate known as the teredo worm, also known as shipworm. In fact, it is not a worm at all, but a burrowing clam. Teredo, a saltwater creature, destroyed $25 million of the bay's wharves and pilings in the first third of the twentieth century. In 1924 the water at the northern end of San Francisco Bay reached its

FIGURE 4.1. Map of Sacramento Delta region.

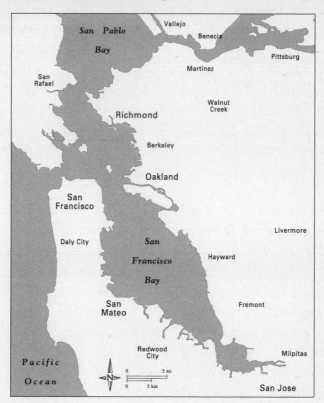

lowest recorded stream flow, and the saltwater content at Pittsburg reached 65 percent. Both Pittsburg and Antioch had used water from Suisun Bay for crops and industry, but in 1926, high salinity forced them to stop.

An idea for a dam to block salt water gained momentum in the 1920s. The U.S. Army Corps of Engineers proposed a saltwater barrier with a moveable crest and multiple locks between Suisun Bay and the junction of the Sacramento and San Joaquin rivers. The corps wanted to create a reservoir, raising the water 5 to 10 feet, but the prospect worried area residents. In 1924, Major U. S. Grant III, the corps' second district engineer, argued that water storage should be the first priority, saying that the barrier's cost could outweigh the benefits. The project was thwarted for numerous reasons, including that the proposed Kennett (Shasta) Dam would not sufficiently curb salinity. California turned to the Bureau of Reclamation, part of the U.S. Department of the Interior, and a major player in deciding how much water moves where and when, for assistance in constructing the planned facilities, and salinity control in the Delta became one of the major goals of the Central Valley Project.[34]

In 1944, Reclamation officials realized the salinity problem in the Delta was more pronounced than they had previously thought.[35] The dam could not entirely control the salinity problem, they believed, which precluded using the Delta as a reservoir as they had once planned. One possible alternative was to build a closed conduit through or around the Delta to carry Sacramento River water directly to the other side without letting it mix with Delta water, foreshadowing later plans for what became known as the Peripheral Canal.

Although most of the Delta was a tidal wetland, the water within the interior remained primarily fresh. However, early explorers reported evidence of saltwater intrusion during the summer months in some years. Dominant vegetation included tules—marsh plants that live in fresh and brackish water. On higher ground, including the numerous natural levees formed by silt deposits, plant life consisted of coarse grasses, willows, blackberry and wild rose thickets, and galleries of oak, sycamore, alder, walnut, and cottonwood. Few traces of this earlier plant life remain; agricultural practices and urbanization have cleared most of it and replaced the natives with invaders.

The prospect of major water exports from the Delta made salt intrusion a primary concern for all water users within the Delta. Various strategies, including saltwater barriers, were considered early on. By the 1930s, a hydraulic barrier, consisting of Delta outflows from upstream reservoirs, was the primary means of salinity control for agricultural and urban water users. Using this approach, both water exporters and water users could agree on the need to keep the Delta's water fresh. The notion of an always-fresh Delta supported by persistent net outflows has endured for more than seventy years, but it is not aging well. This management strategy retains support from Delta water users, but water exporters have come to see increasing risks.

Because of the history of profound and widespread change in the Delta, we are long past the point where the Delta can be "restored" to past conditions, whether it is the Delta that existed before white people arrived or the bucolic agricultural Delta of the last century. No matter what we do, the Delta of the near future will be very different from past manifestations. Its history provides insight into the strenuous efforts Californians have put into finding solutions to collective problems in this pivotal region. And as this history suggests, this process has rarely been simple or smooth, and the solutions have been followed by major investments in the chosen strategy.

In more recent times, as environmental concerns have become central in Delta policy considerations, the search for solutions to the salinity conundrum appears more constrained. Thus, the California-Federal Bay-Delta Program (CALFED) has worked under the premise that the Delta's basic configuration should remain unchanged and that environmental goals could be satisfied simultaneously with those of water exporters and in-Delta interests. Given the crisis of sea level rise and its attendant salinization, now looming in the Delta, it is once again time for California to launch a serious search for solutions. California's problem is the world's problem. This system is central to maintaining perhaps the richest agricultural yield per acre on the planet. When it goes, and it will, food production in California will plummet.

And why this pessimistic viewpoint—"when it goes"? The reason is salt. When freshwater is withdrawn at a faster rate than it can be replenished, the water table lowers, with a resulting decrease in the overall hydrostatic pressure. When this happens near an ocean coastal area, salt water from the ocean intrudes into the freshwater aquifer. The result is that freshwater supplies become contaminated with salt water—as is happening to communities along the Atlantic and Pacific, among many other places on Earth. But nowhere is it more pronounced than in the Delta region of California.

THE SALT WATER FLOODS

Every spring the television news shows the stricken faces of farmers and homeowners whose land and homes are flooded by spring runoff, or a freak storm. In almost every case these floods come from nearby rivers, and thus it is always freshwater that inundates their farms and towns. However dreadful the floods are for people, they have no negative consequence for the land. In fact, such yearly floods were godsends for the early practitioners of nascent agriculture, because in such places as the Nile and Tigris river areas, flooding brought with it new soil and, rather than washing away the soil in the river-fronting fields, left more than it took. But how different it is for the owners of fields and homes that suffer a flood of salt water.

Saltwater floods occur from storm surge, and they really only affect agricultural fields of low elevation fronting the sea. Survivors of the largest storm surges or tsunamis—such as the horrifying event affecting so much of Southeast Asia earlier in this century—had to contend with the effects of salt water washing over the fields and seeping into the soil, or evaporating in the tropical sun in pools and pods that left behind a briny and salt-rich residue. Such events are disastrous for a simple reason: plants, including all agricultural crops, are highly salt intolerant—that is, they grow poorly in the presence of salinity—as we see in Table 4.2.

Although many plants appear to be highly or moderately tolerant, in fact none of them could grow in anything close to the salinity of soil flooded by seawater. Some plants will tolerate higher salt levels in the soil than others (while a few, such as mangroves, actually need a high salt content around their roots), but it does not take much to shut down an otherwise fertile field.

First, high salinity inhibits germination. Adding salt to the soil after a plant's germination mimics a drought. Roots can no longer perform the osmotic functions that transport water and nutrients to areas in the plant that have low concentrations of these needed substances. Salt in soil inhibits the uptake of needed nutrients (which are usually suspended in water), and while perennials as a group seem to do better than annuals, the effect is quite often the same: death of the plant.

Salinization of critical crop areas, such as California's Great Valley, has continued for decades. That means many crops—especially in the San Joaquin Valley, with its uneasy mixing of salt water from the San Francisco Bay and freshwater from the Sacramento and San Joaquin valleys—are growing increasingly tolerant of chloride (one of the two chemical constituents to salt, the other being sodium), as the soils there become ever more polluted with chloride ions. Evolution will probably take a stab at changing plants by increasing chloride tolerance, but this will take many generations and probably can be done more quickly using genetic engineering.

TABLE 4.2. Salt tolerance of various types of plants.[36]

High tolerance	Kochia, sugar beets, altai wildrye, tall wheatgrass, Russian wildrye, slender wheat grass, Siberian salt tree, Sea buckthorn, silver buffaloberry, 6-row barley, safflower, sunflower, 2-row barley, fall rye, winter wheat, spring wheat, birdsfoot trefoil, sweetclover, alfalfa, bromegrass, garden beets, asparagus, spinach, hawthorn, Russian olive, American elm, Siberian elm, villosa lilac, laurel leaf willow, beardless wildrye, fulks altai grass, levonn's alkaligrass, alkali sucatan
Moderate tolerance	Oats, yellow mustard, crested heatgrass, intermediate wheatgrass, tomatoes, broccoli, spreading juniper, poplar, meadow fescue, flax, canola, reed canary grass, cabbage, Ponderosa pine, apple, mountain ash, corn, sweet corn, potatoes, common lilac, Siberian crab apple, Manitoba maple, viburnum
Low tolerance	Timothy, peas, field beans, white Dutch clover, alsike clover, red clover, carrots, onions, strawberries, peas, beans, Colorado blue spruce, rose, Douglas fir, balsam fir, cottonwood, aspen, birch, raspberry, black walnut, dogwood, little-leaved linden, winged euonymus, spirea, larch

While in the past salinization of fields has been caused mainly by the intersection of freshwater with the sea, a second major culprit comes from an unexpected source: salt put on roads to melt snow and ice. Road salt almost immediately makes its way into groundwater, streams, and rivers—and into fields and forests alike. In an effort to keep streets totally clear, more and more cities and towns have also begun to spread salt crystals on their roads in winter. Ordinary sodium chloride is much cheaper than alternatives such as calcium chloride, which is far better tolerated by plants, and it retains the widest use. Every winter some 13 million tons—more than half the salt produced in the United States—is spread on wintry roads. In the six-county Chicago area, the Illinois Department of Transportation alone uses 140,000 tons of sodium chloride during an average winter, and the region's counties and municipalities cumulatively add even more. The figures are similar for other four-season parts of North America.[37]

If any other substance were scattered into the environment in such quantities, there would likely be a public outcry. But for the most part, salt is taken for granted, though it has great and invisible costs, corroding cars and highway bridges. But citizens and politicians have generally accepted those costs as part of the bill for safer winter travel. The environmental costs have generally been taken for granted too. Salt has been used to improve traction since the 1930s, but it was the construction of the interstate highway system in the 1950s and 1960s, and the concurrent boom in car ownership, that dramatically swelled the use of deicing salts. The overall environmental effects of land lost to salinization are one literally subterranean factor in the ultimate loss of agricultural land we will need to help feed the world's billions.

SALT IN THE WATER

The effect of salt on plants is borne out by experience in places other than California. In the 1960s and early 1970s, an uncovered road-salt storage pile sat along the Indiana Toll Road immediately adjacent to Pinhook Bog, part of the Indiana Dunes National Lakeshore. Runoff from the salt pile entered the bog and caused a die-off of bog plants such as tamarack trees, red maples, and sphagnum mosses, which were replaced by more salt-tolerant plants, such as narrow-leaved cattails. Long after the salt pile was removed, the bog plants were only beginning a very slow recovery.[38]

More recently, some of the highest levels of salt in the Illinois–Indiana area have been recorded in artificial wetlands constructed as retention basins. The Conservation Research Institute conducted a study of such wetlands in the Chicago area for the Army Corps of Engineers and found that salt concentrations often reached 650 ppm.[39] Relatively undisturbed natural wetlands in the region typically have concentrations of 8 to 20 ppm; the U.S. Environmental Protection Agency stipulates that drinking water supplies not exceed 250 ppm. Along many roadways, salt-tolerant plants have become more common—annual salt marsh aster, salt meadow grass, narrow-leaved cattails, phragmites reeds, and seaside goldenrod, a diverse array of plants that were mainly new to me. Though no one has been able to precisely demonstrate that salt runoff promotes the growth of these plants, it's clear that they are connected. Salt changes water and soil conditions, and that affects which plants grow where.

More difficult to tease out are the effects of salt on aquatic animals. High salt levels are known to change the composition of invertebrate communities in lakes and springs, but most mammals and fishes are believed to tolerate raised salt levels fairly well. Still, little is known about the effects of particularly high levels of salt in a localized area, such as a vernal pool, the breeding place for frogs or salamanders, which absorb water through their skin. In at least one study, concentrated salt has been shown to inhibit amphibians from crossing forest roads.[40]

And yet, while salting of roads does increase salinity in water and fields, its effect is nothing compared to the effect rising sea levels will have on land. Either way, whether through direct human activity or the encroachment of oceans, agricultural fields are threatened. It will take all of our human resources and cleverness to keep all the human mouths fed in this century and the next.

CHAPTER 5

GREENLAND, ANTARCTICA, AND SEA LEVEL

Greenland, 2215 CE. Carbon dioxide at 1300 ppm.

The geologists scrambled with difficulty over the terminal moraine's high gravel mound, slipping in the debris gouged into the ground by the rapidly retreating glacier. Just over a century earlier, a great sheet of ice had still stood here, seemingly anchored to its foundation. Now all was gravel and scree, with highly grooved and striated bedrock showing only here and there through the mass of conglomeratic gravel and sand. With a practiced swing of the hammer, the senior geologist knocked off a piece of the bedrock. It was sedimentary rock, in fact, very old, from the Paleozoic Era of nearly a half billion years ago. Looking closer, he saw the telltale trilobites, segmented fossils looking a bit like modern-day horseshoe crabs.

Then he corrected himself. The fossil looked like a horseshoe crab from the end of the Cenozoic Era, an arthropod actually increasing in numbers as rising sea level gave its kind ever more habitat. Geologists had always defined their geological units with the fossils found in sedimentary rock, even after the advent of precise radiometric dating. Geology was one of the most historically oriented of sciences, and one still hidebound in tradition. But the world over, geologists informally regarded 2100 as the end of one of the three traditional eras of geological time, the Cenozoic Era—because of the rapid mass extinctions that culminated about 2050. No good name had yet been applied to the new era. People were too busy worrying about staying alive and fed in a world gone madly hot.

The geologist dropped the fossil into the bottom of his bag, tied it off, and labeled it. These were early Ordovician rocks, based on the Scandinavian province Olenid trilobites within. But he was not here for the fossils. He was part of the first geological exploration party ever ferried into a dig site by submarine, under the cloak of the short summer night, when a rich twilight was as dark as the sky got. He bent down and cracked off another hunk of rock. Once again there were numerous trilobites, and on this slab bivalve brachiopods as well. With his hammer he scratched the fossils, all of which were coated in a thick yellow crust.

Everywhere on the fresh outcrop the same yellow crust was present. He put down his sample, shrugged out of his bulky backpack, and took out the most important instrument in his arsenal. Quickly turning it on and calibrating it, he thrust its wandlike probe at the sample. The machine's meter flicked to the right, pegging the maximum reading it was capable of giving. The reading was more than they had hoped for. He twisted another knob, and the audio detector hummed on. A staccato clicking filled the air, like a hundred military drummers lining Red Square. Satisfied, he turned the Geiger counter off. This was a rich uranium prospect, with so much yellowcake present that refining it would be simple. And to think that this find, probably one of the largest previously undiscovered uranium lodes on Earth, had been hidden under ice so recently. He once again fingered the forged passport in his thick pocket. The document would be necessary if they were caught. But he did not think the Danes, who had claimed Greenland centuries ago, would suspect such a brazen geological foray by another country.

The geologist put his equipment back into the pack, and in a loud voice called out to his men to move on. They had many square miles of newly emergent Greenland to explore for more of the valuable uranium and perhaps other heavy elements, which now were so hard to find elsewhere in the world. "Onward," he shouted—in Russian. Let the Danes complain when they finally found out that this part of Greenland had been claimed as a Russian territory. After all, Russia could enforce its rights to this place:

it had wisely retained its arsenal of nuclear weapons. And this was a dog-eat-dog world, with Russia one of the biggest dogs of all.

Antarctica, 2515. CO_2 stabilized at 1500 ppm.

Antarctica. But not an Antarctica that anyone from the twentieth or twenty-first century would immediately recognize. The engineer looked anew at the satellite image. The Antarctic Peninsula was now vastly shortened and separated from the continental mainland—if you could still call it that. The Weddell Sea now extended clear to the South Pole itself, narrowing into long fjords near its terminus. To the west, the Ross Ice Shelf was long gone, as was most of the West Antarctic Ice Sheet. The sea lapped up against the steep inclines of the Trans-Antarctic Mountain Range, now just a long mountainous island because of the flooding of all of Wilkes Land that used to lie above the water level. The island was still a mountain chain, all right, but one now ranging from Beardmore south to Marble Point, a razor-toothed backbone through the new Antarctica—or what was left of it. Most of the unflooded terrain in Antarctica existed only in what was Queen Maud Land—the home of the last continental ice patches on Earth. They were rapidly melting, and that was why this engineer had been shipped down to this godforsaken place.

He studied anew. The last feature on the map was a large freshwater lake, almost large enough to be called a sea, occupying the westernmost regions of Queen Maud Land, filling the depression left by the recently melted ice sheets and glaciers. To the west, this vast depression was bordered by the Shackleton Range. But that was the rub: a huge opening from melting ice in the otherwise wall-like buttress of the Shackleton Range would soon let the lake fill with salt water. Already the rising sea was threatening to flood this growing freshwater lake. The engineer's job was to do something about that. He had been enlisted to construct what would be the earth's largest dam.

Much depended on keeping Lake Hope, as it had been dubbed, filled with freshwater. For the past decade, ship after ship had been conveying topsoil to the huge docking facility constructed at so much cost along the hills that had once formed the western limit of the Amery Ice Shelf. The ice shelf itself, of course, was now part of the Indian Ocean, but the low hills were now the off-loading point for millions of tons of what had been the American Midwest. Thanks to its chronic water shortage and the resulting impossibility of growing crops, the Midwest was now selling its soil—some said its soul too—to the new Antarctic Freehold State. The ships were converted oil tankers, a type of ship found laid up in dry docks everywhere around the globe, because not much oil was left for them to transport. The world needed new agricultural land and there were few places where that could be found. Emergent Antarctica was one such place, and Greenland one of the others, but both faced the same problems. First, after the ice was gone, much of the new land was below sea level and thus invited flooding from the sea unless engineering solutions could be found. Second, there was probably no soil on Earth worse than the regolith left behind by the melting ice in both Greenland and Antarctica. The millions of years of ice cover near the poles caused a lack of plant life necessary to make humus, the organic constituent of soil. The result was a very sterile continent. No amount of artificial fertilizer would change that. Only soil, shipped in at great expense, would make the Antarctic grow.

The engineer sighed. This job would take the rest of his life. Perhaps when he retired he would return to warmer climates. After all, there was no shortage of warm land to bask in. Obtaining water, energy, and food in those places was another matter. But even with the 60-foot rise in sea level to date and carbon dioxide levels at 1,500 ppm, there was indeed plenty of warmth to go around.

POLAR MELTING, RISING SEAS

If carbon dioxide continues to rise, the ice sheets and glaciers of planet Earth will continue to melt. In this chapter, we will look at the source of rising seawater, the great ice bodies on land, and then explore not rates of sea level rise caused by warming, but the singular events that such warming might bring about, such as the above scenarios. Here too, however, we will look at a far less familiar phenomenon, one that only recently has been determined to be a real possibility, and one that more than any other could spell disaster as a result of the most rapid of all kinds of sustained sea level rise—a kind that does not ebb away like storm surge or tsunami. This catastrophe (for that is what it would be) is called ice sheet collapse.

Ice sheet collapse most likely would occur along the West Antarctic Ice Sheet (called the WAIS by specialists), one of the largest reserves of frozen water on Earth.[1] While the alarmists of global warming had long used the specter of its demise as a scare tactic, they took some comfort from the fact that no one had ever actually demonstrated that ice sheet collapse could in fact take place. All that changed in 2009 with the discovery that just such an event had happened to the West Antarctic Ice Sheet during a period of rapid global warming and ice melting that occurred about a million years ago.[2] Ice sheet collapse went from a hypothetical alarm to an all-too-real prospect. Exploring the formation, melting, and inevitable collapse of ice sheets (inevitable in the sense that over the long run, it has happened and happened again, but also inevitable in the short run—the next millennium—if we let CO_2 rise above 1,000 ppm and stay there), we will be better able to confront the possible fates of Greenland and Antarctica. All estimates of sea level rise depend on our confidence that if the ice sheets indeed continue to melt, they will turn to water in some regular and predictable fashion.

While the slow but accelerating rise in greenhouse gases would seemingly be the ultimate arbiter of our planetary fate, much depends on the rate and method of ice sheet disappearance. The greatest danger comes if Greenland's ice sheets vanish;[3] we would suffer catastrophic sea level rise. But that would take a long time. More ominous—and immediate— is the possibility that a large part of the West Antarctic Ice Sheet could disintegrate over just a decade's time, leading to a 15-foot rise in global sea level within perhaps twenty years of the initiation of this spectacular event—one that will be a show for the penguins, seals, and tourist cruise ships but a nightmare for the world. But before tackling the question of whether they will entirely melt, let us first look at just what an ice sheet is, and at the factors controlling their growth or diminishment.

ICE CAPS AND ICE SHEETS

There is often confusion about the difference between ice sheets and ice caps.[4] Ice sheets are larger than about 20,000 square miles, whereas ice caps are smaller. Ice sheets sit atop land. There is certainly ice that floats in water, but as we saw earlier, such ice is sea level neutral. If it expands or disappears (as it is rapidly doing in the Arctic oceans now), there is no effect on sea level. Not so the ice sheets. If they melt, the water that composes them pours into the oceans, either immediately (if the sheet extends to the coastline) or eventually, through discharge into rivers (which is happening now at ever-faster rates in the northern Atlantic region).

How do ice sheets form in the first place? To build an ice sheet, snow must accumulate to such an extent that the winter accumulation is greater than the winter melting (known as ablation), because the seasons dictate a cyclic pulsing of growth and destruction. Much is now known about this process. When ablation exactly balances growth, the ice cap is said to be at equilibrium. Because summer (with sunlight for most of the day) and winter (dark most of the day) are so different, the ice caps show greatest accumulation in winter, when temperatures are lowest, and highest ablation in summer. The net rate of growth appears to be very slow. If conditions are favorable for accumulation, it can still take 10,000 years for a mile-thick ice cap to form, a rate of about 10 inches per year. But conditions were rarely favorable in the past, as the global climate commonly showed rapid changes in temperature, sometimes of as much as ten degrees in as little as ten to twenty years.

It is not just the piling on of snow and ice that builds an icecap. Ice is heavy, and as the great mounds build up on the continents, they commonly cause the land under them to subside, further slowing accumulation. Just the opposite happens if and when the ice cap melts—the land springs upward in what is called an isostatic rebound.

Ice sheets grow vertically, just as sedimentary rocks do.[5] But unlike sedimentary rocks, ice sheets are affected by altitude and the angle of sun hitting them, known as insolation. More ice accumulates at higher altitudes, because such regions are colder than land at sea level. In addition, the accumulation of ice produces what is known as a positive feedback: if the ice sheet grows at its edges, as well as upward, the net albedo rises—that is, the reflectivity of the planet increases, which lowers global temperatures. Better than any other sedimentary cover, the white ice and snow reflect sunlight back into space. That makes the global mean temperature drop a small amount, causing a slight increase in the rate of accumulation.

The rates and character of accumulation and ablation differ greatly. During the winter, ever-colder temperatures cause accumulation that remains steady at rates of about 6 inches per year. Ablation, however, shows a very different pattern. There is no net ablation at temperatures below 1 degree Fahrenheit, but as soon as that temperature is reached, ablation shoots up rapidly, with greater than 13 feet of ice lost per year simply with the rise from 15 to 32 degrees Fahrenheit. The temperature of about 23 degrees Fahrenheit seems to be the tipping point from accumulation to ablation.[6] All of this seems odd, because as we know, a far higher temperature is normally required to melt ice. But an ice sheet is not an ice cube, and other forces are at work. Nevertheless, temperature is clearly the main factor deterring whether an ice cap grows, remains at equilibrium, or shows net ablation for the year. But what controls temperature? Herein lie the crucial elements: two main factors deciding temperature are the amount of sunlight hitting our Earth, and the concentration of greenhouse gases.

The melting or freezing of large ice masses has a surprising effect on global temperature. Whereas ice reflects sunlight, water absorbs it. That means that periods of cold, when ice sheets are growing, produce a positive feedback in temperature. This happens as follows: as ice melts, rocks are exposed that warm faster than ice does to sunshine, and the world warms. Then ice that previously was just cold enough to stay frozen begins to melt, exposing more rocks, and on and on, growing ever warmer. The colder it gets, the colder it gets, because more ice reflects more sunlight back into space, thus cooling the planet. New studies of the mechanisms of glacial retreat along the coast of Greenland—the very places where a melting ice sheet most quickly disappears—suggest that rising sea level might be producing its own kind of positive feedback, with more water causing greater heat absorption, which melts more ice and causes ever-higher sea level.

This phenomenon makes sense in the context of ice shelf and ice sheet thinning—where an ice sheet is entirely on land, and an ice shelf is a great block of floating ice, kind of like a super iceberg, that may or may not be resting next to land (they often "run aground" in storms)—a precursor to melting and disintegration. It appears therefore that glacier or ice shelf thinning is the key preconditioning factor for collapse, retreat, and accelerated melting. The mechanisms for ice shelf thinning include basal melting (from warming ocean waters), surface melting, and reduction in glacier inflow—that is, new ice coming down the glacier that wedges under existing ice. These mechanisms lead the ice shelves to eventually collapse.

GREENLAND, TURNING GREEN

To anyone flying the great semicircular route from London to North America's West Coast, Greenland is a familiar sight. Departing London in the late morning, by mid-afternoon you are flying over the massive island's vast, white, desolate landscape, with barely a shred of green visible. All is snow, ice, and rock. It ought to be white and desolate; its northern tip lies within 450 miles of the North Pole. The seas and waterways bounding Greenland—the North Atlantic Ocean, the Arctic Ocean, Davis Strait, and the Denmark Strait—are all frigid places where icebergs are common. The land surface covers nearly a million square miles—and is 85 percent covered with ice sheet.

The land itself has a peculiar cast to its geography. The coastline is made up of seemingly impenetrable mountains, bisected perpendicular to the coast by innumerable fjords, most of which have glaciers protruding at some place within their confines. It is as if a thousand drainpipes jutted out of a giant bowl of melting ice, each a culvert sluicing meltwater into the sea.

Flora and fauna are challenged by the extreme climate, with winter temperatures dipping as low as –76 degrees Fahrenheit. Summers are much warmer, with maximum temperatures reaching as high as 75 degrees Fahrenheit,[7] but they are ephemeral, with the darkness of winter and the nearly daylong night of the extreme high latitudes held in impatient abeyance.[8] Nevertheless, there is enough light and warmth in the brief summer to allow a small flora to exist, as well as native mammals and birds, such as musk ox, wolf, lemming, and reindeer. Almost all of these animals and plants live on a thin coastal fringe on the extreme southwestern coast of the giant island, by far the warmest region of Greenland. There are humans on Greenland as well, and they too mostly occupy the southern coast. Most are Inuit, current descendants of the first, hardy, often nomadic inhabitants who lived off this land for many millennia, but there also have been Danes ever since Denmark annexed Greenland long ago as essentially a colony.[9] Fishing is the major industry; agriculture is possible on only 1 percent of Greenland's surface.

There is no doubt that the Greenland ice sheets (unlike those in Antarctica, Greenland's are not formally named, and thus they are lumped together here) are already melting, but how quickly, and in what fashion, remains unknown—but not unobserved.[10] So many variables, so little time to even identify them all. The large ice sheets of Greenland and Antarctica are like icicles hanging over our heads. If the ice sheets stay frozen, or even if they stay mostly frozen, melting only slightly and slowly, then this book will be irrelevant and quite wrong in many predictions. Earth will experience very slow sea level rise combined with whatever thermal expansion of the ocean already has in store for us, and our children's children will not be dealing with the economic burden of building new ports and airports on higher ground. But at what rate does melting become dangerous to humanity? And would melting of the ice sheets be irreversible? Once melted, how much would CO_2 levels have to drop before the ice sheets formed again?

At 1 million cubic miles in volume, up to almost 2 miles thick, and a little smaller than Mexico, the Greenland ice sheet holds a tremendous amount of freshwater. If it totally melted, it would raise global sea level by about 22 feet (this figure is formally referenced later in this chapter). How long it might take to disappear is a matter of much controversy and is among the most critical of the questions pertaining to the rate and magnitude of sea level rise. Some estimates say it will totally melt within one thousand to several thousands of years—if melting were the only mechanism by which it lost mass—depending on the magnitude of future warming. But other factors could cause it to vanish more quickly. While in any event melting causes the ice to convert to water, the rates of melting really depend on whether it melts from the top down or breaks apart from the bottom up. Disintegration from the bottom up could lead to sea level rise rates far higher than those of simple top-down melting.

New observations from 2007 bear on the question of Greenland ice sheet melting.[11] (It is interesting and alarming that there seems to be little argument that it is indeed melting—the contention is over the mechanisms.) The velocities of several large glaciers "draining" from the Greenland ice sheet to the sea, already among the fastest flowing on Earth, have recently doubled to reach about 7 miles per year. The draining action is similar to that of a river coming out of mountains, only in this case the water movement is in ice, rather than liquid water. In addition, the ice sheet experienced a greater area of surface melting in 2007 than at any time since systematic satellite monitoring began in 1979.

Both of these changes—from-top-melting and ice flow into the sea by glaciers—increase the mass that is being lost from the ice sheet, with the implication that current (but conservative) estimates of global sea level rise over the next century—of about 2 feet at most predicted by the Fourth IPCC—are surely underestimates. James Hansen has produced the highest estimate of all: that by 2100 there will be more than a 6-foot rise, perhaps as much as 15 feet.[12] As we've seen, like all ice sheets, Greenland's gains mass through snowfall and loses it by surface melting and runoff to the sea, together with the production of icebergs and melting at the base of its floating ice tongues. The difference between these gains and losses is called the mass balance; a negative balance contributes to global sea-level rise and a positive balance keeps water levels closer to normal. About half of the discharge from the ice sheet is through twelve fast-flowing outlet glaciers, most no more than 6 to 12 miles across at their seaward margin, and each fed from a large interior basin of about 19,000 to 39,000 square miles. The behavior of these dozen glaciers dictates the ultimate mass balance, and their sensitivity to current global warming is what is pertinent.

Two changes to these glaciers have been observed in 2007 and 2008 that may relate to the fate of the Greenland ice sheet. First, the floating tongues or ice shelves of several outlet glaciers—each several hundred feet thick and extending up to tens of miles beyond the grounded glaciers—have broken up in the past few years.[13] Second, measurements of ice velocity made with satellite radar interferometer methods have demonstrated that the glaciers' flow rates have approximately doubled over the past five years or so.[14] The effect has been to discharge more ice and increase the mass deficit of the ice sheet from a little more than 12 cubic miles a year to in excess of 36 cubic miles per year. Two of the larger outlet glaciers, Jakobshavn Isbrae in the west and Kangerdlugssuaq Glacier in the east, are all south of 70 degrees north, and because of this relatively southern latitude compared to Greenland as a whole, they may be more influenced by warming than the other ten—suggesting that glacier discharge and changing climate may somehow be linked. Satellite data from passive microwave instruments show that there has been a marked increase in the area of Greenland affected by summer melting, as well as in the length of the melt season. Indeed, over twenty-seven years of observation by satellite, 2002 and 2005 are records for the extent of melting.[15]

It is just the global warming that might be causing the increased melting of one of the largest of all of Greenland's glaciers, a forbidding river of ice named the Jakobshavn Isbrae glacier. Comprising a full 7 percent of the entire Greenland ice sheet (by area), Jakobshavn Isbrae transports huge volumes of ice inland to the head of a large fjord, which eventually exits into Baffin Bay. A current of relatively warm water travels as an ocean current within the vast reaches of Baffin Bay, so any ice calving off the Jakobshavn Isbrae glacier, falling into the fjord, and traveling into Baffin Bay soon melts—adding more water to a rising sea. Scientists measuring the rate of ice flow down the vast Jakobshavn Isbrae have noted a vast increase in this rate over the past decade.[16] Associated with the rise has been a measurable thinning of the ice itself as it approaches the sea. Were the faster ice flow and thinning ice two related phenomena? For a while, it was thought that the increase in speed could be explained by reduced friction along the base of the glacier due to ever more meltwater. This reason makes intuitive sense: ever-greater meltwater would lubricate the glaciers' contact with underlying bedrock, allowing increased speed. But new research shows that some other factor might be at work. The very rate at which the glacier's ice reaches the sea—with great hunks calving and falling or being pushed into the sea to float off as icebergs—might be involved as well. The thinner ice calves more quickly, and as it is carried into the sea, it makes room for the glacial ice behind it to reach the water's edge. The ice is getting thinner because ever-warmer air causes the entire glacier to become a shorter column of ice—the glacier thins from above. But a second aspect of the thinning is less apparent but is now becoming clearer: seawater infiltrating at the base of the glacier causes melting along the bottom. This process occurs not just with this particular glacier in Greenland. Recently several great ice bodies in Antarctica, along the Wordie, Mueller, Hones, Larsen A, and Larsen B ice shelves, have "collapsed"—that is, lost integrity and fallen piecemeal into the sea, where they eventually melt.[17] Both in Greenland and in Antarctica, the oceans' newly warmer water has increased the rate of melting of the ice sheet bases, causing them to thin from below just as the warmer air is simultaneously causing thinning from above. The glaciers calve more, and the rate of movement increases. But the negative feedback loop comes as sea level rises, which causes the thinning effect to migrate up the glacier bases ever farther inland and induces more rapid thinning and melting. Higher sea level causes higher sea level. These and other signs of enhanced activity along the margins of the Greenland ice sheet are ominous.

To climatologist Richard Alley, one of the world's experts on ice sheets and the climate of previous eras, this increased activity may change how we estimate sea level change in two ways. First, Alley noted that current models of melting and subsequent sea level change might underestimate the ice's melting behavior and rates. New mathematical models that calculate estimates of temperature change and ice melt are just beginning to include the newly observed melting rates, but if the new data prove important, then sea-level projections may need to be revised upward.[18] Second, because the areas where the vast conveyor currents in today's oceans sink to the bottom, carrying oxygenated water with this vast downward spigot of ocean water, occur immediately adjacent to Greenland, any notable increase in freshwater fluxes from these ice sheets may induce changes in ocean heat transport and thus climate. In the horrible 2004 film *The Day After Tomorrow*, the start of all the mayhem was when this same current shut off. Unfortunately, the movie got things backward. Although Europe might cool for a while, the conveyor currents have in the past turned off due to warming. I will revisit these in greater detail in Chapter 7, which discusses greenhouse extinctions.

Scientists have used many models to try to figure out the real "lag effects" of the current greenhouse gas increases (where lag effect is further warming by CO_2 and other greenhouse gases already in the atmosphere, but not yet fully warming the planet), added to projected temperature increases from a variety of future greenhouse gas level scenarios—cases where the level of greenhouse gases are kept the same, then raised in increments to estimate future temperatures. The 2002 IPCC estimated that the danger level for wholesale melting is only about 3 degrees greater than current temperature. If the balance between annual melting and accumulation tips conclusively toward melting, then ice sheets will start to make the sea level rise.

When might the switch happen from net accumulation to melting? Some models say that the changeover—and the rise of temperature of about 3 or 4 degrees—might not happen until 2350. But then other models show it happening by 2100. In either case, it looks like Greenland will reveal its rocky surface for the first time in thousands of years. The models suggest not if, but when. There is some optimism for the near term, because other estimates show that complete loss of the Greenland ice sheet will take 1,000 years to complete, ending with a 23-foot sea level rise. But Greenland would not be alone. Any temperature rise that causes the complete collapse of the island's ice would be melting even larger ice sheets as well, most importantly the West Antarctic Ice Sheet. In the long term seas could rise even more than 23 feet, although a 2009 report halves that estimate to 11 feet.[19] However, even that would be disastrous.

An interesting caution from the models was the unexpected finding that once the ice disappears it may be impossible to reconstitute[20]—unless the temperature drops so extremely that a new and major continental ice age would arise as well. As long as there are hydrocarbons to burn, and that over the next 1,000 years less-developed countries keep burning the seemingly limitless coal deposits (with oil long gone), a melted Greenland would not arise again re-iced.

NOT ICE BUT OIL

Greenland has challenged human occupiers since almost 5,000 years ago. By that time virtually every other place on the globe (except Antarctica and the more isolated island groups) had been visited and colonized by humanity. European civilization made its first foothold in the Middle

Ages,[21] around the year 982 BCE, with the arrival of Erik the Red, a re-sourceful and ambitious Norwegian who led an expedition in small open boats across the perilous seas from Scandinavia and established a colony. At first it prospered, because the earth had entered one of its warm phases. But by the middle of the 1400s, the earth had returned to a deep freeze, and the last of the hardy Viking settlers withdrew, leaving behind a cold, infertile landscape. The hardy Inuits watched them go, surely with mixed emotions.

Despite the fact that Norwegians were the first European colonists, Denmark first staked a claim on Greenland when missionaries con-structed a trading post on the coast. In the 1800s there was a series of claims and counterclaims, even by the U.S. government, but Danish au-thority held sway. Today it is an important military and meteorological outpost for NATO forces, and the giant Thule Air Force Base has great strategic importance, situated as it is on the great circular bombing run between the United States and Russia. Over the past two decades Green-landers have made ever more hue and cry for independence. The native settlers want to break from Danish rule, while the Danes are under-standably hesitant to give the place up, especially since they have been pumping millions of dollars per year into its fragile economy.

The question of who owns Greenland will become increasingly im-portant over the next decades, as more and more of its ice sheet melts, re-vealing what lies beneath. Even with its extensive cover, geologists have discovered economically important deposits of lead, zinc, and aluminum, and they expect other metals as well. Perhaps most valuable could be vast new petroleum reserves. Already by 2006, exploratory wells hit pay dirt, and the Greenland government, now a quasi-autonomous Danish protec-torate, has since allowed major oil companies to undertake new searches in ever more remote regions.[22] It is extremely expensive to sink wells over ice-covered regions, and some places will not support rigs at all. But what-ever the other ramifications, a Greenland free of ice could become a new economic mecca, especially since the price for a barrel of oil near the end of this century, and into the next, will surely be far higher than it is today. Greenland's oil should come on line just as many of the currently viable oil fields in the Middle East, South America, Indonesia, Africa, and North America are winding down, their product already gone up in smoke. This drop from what currently might be peak oil is causing alarm among the traditional oil producers, as well as the new ones, such as Russia—which means covetous eyes are now turned toward Greenland, and even more so to deeper water off its shore, areas currently in international waters. Canada, Russia, the United States, Britain, and several Scandinavian countries are all trying to make the case that some of the high Arctic seabed belongs to their particular territory because of geological history. Many of the claims are dubious at best and would be laughable if not for the gigantic, nuclear-armed militaries some of the claimants hold.

THE TIMING OF GREENLAND'S GLACIAL MELTING

When will the ice sheets go if temperatures keep rising? At what temperatures will this conversion of ice to freshwater flowing into seawater take place in earnest? Luckily, one of the strongest parts of the IPCC work is in predicting future temperatures. But in fact the hardest part is correctly deducing how various rising temperatures will impact the ice sheets. We know that if civilization stopped emitting greenhouse gases right now there would still be a decade or more of continual global warming. But I have not found any report stating that our current temperature levels would cause the ice sheets to disappear. However, a new model estimates the fate of the Greenland ice sheet under various possible temperature regimes.

These new estimates come in the 2004 article published in the prestigious scientific journal *Nature*, by Jonathan Gregory and two colleagues. Right at the start of their "brief communication," as the journal calls their work, they state that the Greenland ice sheet will be eliminated except for a few holdover mountain glaciers in the region if global temperatures become a little more than 5 degrees Fahrenheit higher than they are today. In Chapter 2 we discussed various scenarios for temperature increase, and most of them indicate that temperatures will far exceed the 5-degree benchmark in a relatively short time. But how short a time?

Like so much on this planet, trade-offs will come with future warming. More warming of Greenland will also lead to more precipitation, according to this article. If such precipitation occurred now, as snowfall, the Greenland ice sheet would actually grow, and would continue to do so at temperatures up to almost 5 degrees Fahrenheit warmer than now. But temperatures above this level cause melting of the new ice to exceed precipitation, and if the temperature climbs higher this melting rate increases as well. If temperatures increase to almost 13 degrees Fahrenheit warmer, which is not impossible, according to the 2007 IPCC, the entire ice sheet will melt away in less than a millennium. But might this happen at lower temperatures? If so, perhaps Greenland's ice sheet melting will occur far earlier than 1,000 years from now. The result of the loss of the Greenland ice sheet is well known: again, about a 22-foot rise in sea level.[23] Moreover, there would be no lag time—it is not as if the ice sheet goes away but sea level takes decades to rise, similar to the lag time of thermal expansion of the oceans.

In their model, Gregory and his colleagues ran mathematical simulations of climate under a variety of atmospheric CO_2 concentrations. These models accept the premise that Greenland will experience greater temperature rise than the planet's midlatitude and tropical regions under the same CO_2 concentration, a principle all climate modelers accept. The CO_2 levels inputted into the model are estimated to occur based on the 2007 IPCC report. Because the IPCC uses a variety of CO_2 concentrations, in order to look at the best and worst cases of still-possible CO_2 values, a graph of these findings will necessarily have a lot of potential error, and one such model result is shown in Figure 5.1. The premise is that CO_2 concentration reaches a certain level and then stabilizes into a constant. Boston used seven CO_2 concentrations: a minimum of 450, then 550, 650, and 750, ending with a maximum of 1,000 ppm.

Although the CO_2 levels in the model indeed could occur—450 ppm certainly will, and, from all I see, each of the others will be not only attained, but exceeded—Gregory and his team used a very conservative time line. They put the expected date for the highest level, 1,000 ppm, in 2375, but most other models show the planet reaching higher levels sooner than that. Nevertheless, the results of the modeling are the same no matter when they occur.

The inevitable atmospheric level of 450 ppm will result in the loss of Greenland's ice after a millennium passes. This finding alone tells us that the 450 ppm mark, often used by modelers and seers of global warming as the last stop before the "tipping point" for all kinds of nasty, warm sur-

FIGURE 5.1. Greenland and its ice sheet.

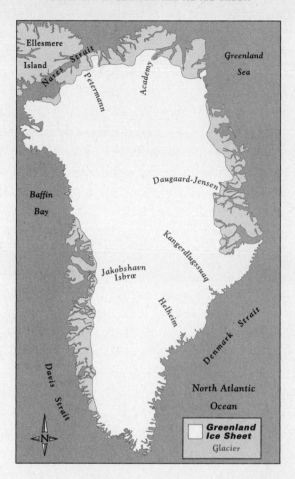

prises, is in fact very dangerous after all. The message is that human civilization cannot let CO_2 levels stay at this point and that we must do all we can to ensure that CO_2 never gets that high in the first place. And since other greenhouse gases besides CO_2 are warming the atmosphere, CO_2 concentrations of 450 ppm will have even more impact, because of methane and water vapor that increase in the atmosphere as temperature rises. These are both potent greenhouse gases, as we have seen. They will add to the warming but are rarely factored into the various models.

At even higher levels of CO_2—750 ppm and 1000 ppm—the tipping point of 5 degrees Fahrenheit is reached in about thirty years, around 2040. Such a rise means that in a century the temperature of Greenland will be a whopping 13 degrees Fahrenheit higher than now! There is no way the rate of sea level rise will remain low under this circumstance, because the same factors that warm Greenland will also cause warming of Antarctica, and it is probable that Antarctica's ice sheets would melt simultaneously with the Greenland ice sheet. Within a century or two the result would be catastrophic flooding of 10 feet or more.

The final result of this modeling is itself shocking. Let's assume that we humans decide at some point that we want to be in this living business for the long run. After failing to save the Greenland ice sheet (and parts of the Antarctic ice sheet), we finally switch to non-carbon energy sources and engineer the current superabundance of CO_2 in the atmosphere (but we cannot get rid of it all—that result would doom plants and, ultimately, us). It begins to snow again in Greenland, but no ice sheet forms. According to the Gregory paper, the loss is irreversible.

THE MELTING OF ANTARCTICA

Voluminous as it is, the quantity of freshwater tied up in the Greenland ice sheet is utterly dwarfed by two ice sheets covering large areas of Antarctica, designated the Western and Eastern Antarctic ice sheets. In 2007 and 2008 there were reports that, unlike the Greenland ice sheet, universally acknowledged to be shrinking through melting, the West Antarctic Ice Sheet is actually increasing in size.[24] The growth is occurring in the areas farthest from the sea and is caused by an excess of snowfall compared to melting in the ice sheet's cold interior—although deep down there is melting, this is more than compensated for by the amount of snow on top. But the rate of melting along the edges is very poorly constrained, even worse than in Greenland.

Early in this book I recounted my experiences there. Antarctica is always pictured as a forbidding place of ice and cold—the most remote place on Earth, the most extreme climate, and the one continent without native humans or non-marine mammals. It is a far more austere environment than even Greenland, but there are political differences between Greenland and Antarctica that may ultimately prove to be as important for the fate of sea level change as the physical differences. Because it is the sole continent that never harbored indigenous populations of humans, Antarctica has always been a curious thorn in the side of world

FIGURE 5.2. Map of Antarctica, showing what is known about bedrock geology. All the rest is where ice now covers the continent, hiding any possible mineral/oil reserves.

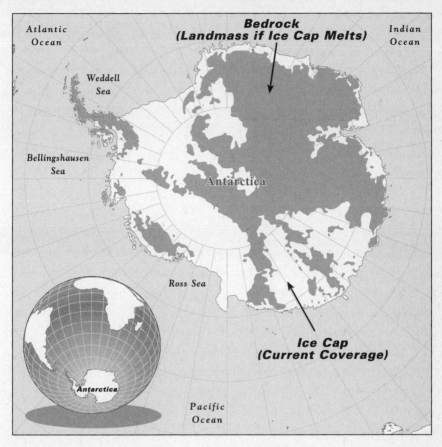

politics. To ensure a political presence, many nations keep permanent bases there largely through scientific work, although the bases themselves are maintained by various militaries.

In the past, the main value of Antarctica came not from its land but from its surrounding seas. Unbelievably rich in plankton and surface productivity, this water thrives with larger animals, including the largest whale population on the planet. Because of this, Antarctica for two centuries was a prime locale for whaling and would be still but for the partial protection afforded by the international whaling treaty. But this treaty is part chimera. For example, Japan flouts a similar international agreement by taking large numbers of whales, supposedly for "scientific" purposes. Yet at the same time, that country still sells that whale meat as food for dogs as well as for humans. Using the same logic, any number of countries could begin removing minerals or oil in the name of science from Antarctica. It is what lies beneath the ice, as well as the continental shelves surrounding the icy continent, that imperils the current "hands off" protocol for Antarctica.

The first major treaty laying out a series of ground rules was the original Antarctic treaty, signed in 1959 (but not taking effect until 1961) by the nations geographically closest to Antarctica: Argentina, Australia, New Zealand, and Chile—each believing that their proximity gave them more of a say in ruling the continent than those countries farther away.[25] But others involved were Great Britain, America, and the Soviet Union—the big boys, and big bullies, on the international block. Each of these countries maintained permanent military-run bases in Antarctica. Ostensibly this military presence was peaceful, but surely some kinds of weapons are stockpiled in the bases, "just in case." By the late 1900s the various countries with current stakes in the continent realized that any new treaty would have to deal forcefully with future mining and oil extraction. Thus, in April 1997 the Madrid Treaty was signed, ostensibly to augment the Antarctica Treaty. Its major proponents, Australia and New Zealand perhaps most strongly, put in place a ban on mining and oil exploration for fifty years, or until 2047. Also prohibited were dumping or storing toxic waste in Antarctica, testing weapons (Antarctica would be ideal for testing nuclear arms), and destroying flora and fauna.

Surely a sign of the future political upheaval awaiting the continent, ratifying the treaty was anything but easy. An unusual aspect of the pact was that all parties must agree to make any future decisions by consensus. Furthermore, for the treaty to be valid, twenty-six of the fifty countries involved in negotiations were required to sign on this newest ban on mining, and by 2003 the treaty still lacked two countries necessary to ratify, Russia and Japan. Even so, the treaty went into effect in 1998. As of 2009, of the forty-eight signatories, twenty-eight hold decisive votes, while seven members claim ownership to small parts of the continent.

The fifty-year ban on mining is tenuous at best. Because of concerted opposition from interested parties with an eye toward exploiting Antarctica's natural resources, the treaty contains a "walkaway" clause that allows any signatory to withdraw from the treaty and begin doing what the agreement specifically outlaws: mining. Although Antarctica is economically valuable today mainly for its tourism, it is the continent's metals and oil that will lure industrialized countries to it in the future—and its riches could have dire consequences for the world.

HOW ANTARCTICA GOT ITS ICE SHEETS— AND HOW IT COULD LOSE THEM

Continents are the eternally joyriding bumper cars of the earth's crust; unlike the more sedate oceanic crust, continents spend tens of millions of years drifting one way, only to smack into another continent and run off in a new direction. Longitude and latitude have no long-term meaning for continents; their tropics become the high-latitude cold regions, and then back again. And so it has been for Antarctica. It has been cold, then warm, and back again numerous times. But for understanding our future climate, the more recent past of Antarctica—that is, in the millions of years since dinosaurs roamed the earth—holds great clues. For instance, just when did Antarctica get its great continental ice covers, the thick, freshwater ice sheets that hold the fate of sea level—and civilization, not to put too fine a point on it—in their (still) icy grasp? When did that ice form, and what was the geography and atmosphere of Antarctica like at that time?

Although the overall pattern of Antarctica's geographic movement—known as continental drift, and driven by plate tectonics—has been understood for some decades thanks to data coming from paleomagnetic investigations, until recently much less was known about the history of its ice. Happily, that is changing. The first thick ice formation has been estimated to have occurred perhaps 30 million to 40 million years ago, about midway between the end of the dinosaurs and the present, during the old Tertiary Period (now known as the Paleogene Period).[26] But a second crucial study had a far more important bit of information—not just the *when*, but the far more crucial *why*? Why did ice begin to form then?

The accepted knowledge has been that the ice began to build up in Antarctica tens of millions of years ago because of a radical change in ocean currents around the continent that, according to prevailing theory, flipped on the refrigeration switch. This idea emerged in the 1970s,[27] when oceanographers pioneered new methods of determining ancient temperature information from the oxygen isotopic study of ancient deep-sea cores and identified what appeared to be a progressive global cooling over the past 65 million years that included several steplike decreases in temperature at both 34 million and 15 million years ago.[28] These two significant temperature drops, superimposed on a longer-term and more gradual drop in global temperatures, were interpreted to have been the result of tectonic movements of continents, which through their new configurations forced many ocean currents to change their positions, flow, and patterns. That major current patterns can influence climate along the coast of a continent has never been in doubt, and it made good sense that that separation of Australia and South America during the Tertiary Period—one of the last acts of the long-term disassembly of the ancient supercontinent of Gondwanaland—would have radically changed currents in the southern oceans. The separating continents would have created new and enlarging ocean gateways between the previously connected landmasses, corridors through which new ocean currents would have begun to flow, affecting the regional climate through warming, cooling, or both.[29]

The new currents caused new east–west flow around the Antarctic continent. Where before warm waters coming from the north had bathed that continent, now it was encircled by water that never experienced the warm sun of lower latitudes. According to this idea,[30] Antarctica was surrounded by currents of ever-cooling water and the stormy, cold, gray waters of the southern ocean around Antarctica became thermally isolated from the warmer oceans, and cooled—and cooled some more, eventually becoming covered with ice.

For nearly thirty years all were satisfied with what became known as the Shackleton and Kennett hypothesis.[31] Other possible influences on Antarctica's cold climate—such as the level of greenhouse gases, or orbital aspects of the earth that would affect the amount of solar energy hitting the planet—were dismissed or downgraded. Those long-ignored influences would be essential not just to determining what had made Antarctica cold but to understanding the elements that would affect our climate in the future. Scientists would have to assess the relative importance of atmosphere, ocean, continental position, and solar flux and its effect on the earth's orbit. Such a holistic evaluation finally occurred in 2003, when climatologists Robert M. DeConto and David Pollard took a new look at Antarctica's cooling and its ice cap formation.[32] Unlike their scientific forebears, they did not brave the far southern seas to extract difficult-to-obtain sediment cores from the ocean bottom and then analyze them in multimillion-dollar laboratories. Instead they sat in warm labs, inputting numbers into standard-issue personal computers—and came up with new conclusions about the cooling of Antarctica. Their bottom line: it was declining CO_2, not changing ocean currents, that had made the continent so cold.

Part of what led climatologists to look anew at Antarctica at the start of the new century was a sensational event that occurred in 2002, when part of the Larsen B Ice Shelf, partly on land, partly a frozen seawater region abutting the land surface of Antarctica, broke off and fragmented.[33] The area of ice that disintegrated into an armada of icebergs was about the size of Rhode Island. The sensational aspect came from the speed at which it happened, and the sheer magnitude of the ice affected: the ice covering the Larsen B region is about 600 feet thick and has been in place for at least 12,000 years. But recent thinning apparently made it fragile, and the anomalous warm summer of 2002 seemed to do the trick. Off into the ocean it went.

THE RATE OF CONTINENTAL ICE MELTING

The pioneering late eighteenth-century geologists Charles Lyell and James Hutton proposed the Principle of Uniformitarianism to explain the processes that resulted in the earth's current rock record.[34] They held that processes of the past were the same as they are today. Ice sheets are made up of sedimentary beds composed of ice rather than rock, but they still should obey the principle that the present is the key to the past, and should obey principles of sedimentary rocks, such as lower strata being older. But in some fashions the ice sheets do not act like sedimentary beds; although sedimentary beds are subject to erosion and can be completely ground to dust if rock accumulation ceases, ice sheets can both slowly erode and more quickly disintegrate. We know from the rock record that ice sheets can and do disappear on a scale of mere centuries, not the tens of millennia that seem to affect rock erosion. Yet while the headlines for the past two years have concentrated on the rate of pack ice appearing (that is, seawater ice floating over oceans in high latitudes), the reality is that all the ice cover of the North Pole region could melt and not cause a fraction of an inch of change in sea level from the melting itself. (As we've seen, of course, the melting would have a profound ancillary effect on the heat of the planet, because the loss of the ice decreases albedo, thereby further heating the planet, thereby melting more ice—a positive feedback mechanism.) Nevertheless, it is not this kind of melting that affects sea level estimates. For this we need to know the rate of continent ice sheet disintegration, which leads to melting.

Ice sheets are made in ways fundamentally different than how they are unmade. An ice sheet grows through the formation of ever more solid ice. But the melting of an ice cap is a wet process that can occur rapidly—far more rapidly, in fact, than ice sheet growth. The rapid rates come from the way the melting occurs.

While most study of ice sheet history has been done for the recent ice ages, which began some 2.5 million years ago, other scientists have looked for evidence of ice sheets in deeper time. There is now a good understanding of ice sheet activity over the past 55 million years.[35] If we went back to that distant time, during the Eocene Epoch of the Paleogene Period, we would find a planet with no ice sheets at all. But major tectonic events as well as plate tectonic placing of continents conspired to cause a long-term drop in global temperatures. As our planet cooled, ice sheets grew in both the Northern and Southern hemispheres.

All of the past 65 million years—in fact, for the past 200 million years—has been set against slowly declining carbon dioxide values. For the past 50 million years the poles have been increasingly cold, and over the past 40 million years they have had ever more ice. Our time is a reversal of that. Although several other brief warm episodes have occurred over the past million years, such as the one that brought sea level in Florida and elsewhere up by 10 feet more than 100,000 years ago, all in all it has been a long-term cooling. Now we are faced with not only warming but also melting, and not slow melting but rapid melting.

EMERGING GREENLAND AND ANTARCTICA

What will happen when the ice melts on Greenland and Antarctica? First, it will spawn ever more icebergs, with vast quantities of water exported away from the landmasses as on great ships. But as the ice melts inland, away from the warm ocean toward the interior where the coldest temperatures reside, melting will produce enormous volumes of freshwater. The Northern Hemisphere subcontinent, as well as the Southern Hemisphere true continent, will rise up out of the ice, reminding us it is land. But that land will not last long. The melting of the ice sheets will cause sea level to rise rapidly.

The first to go—in fact already well along in disappearing—is the Arctic ice cap. In 2007 an astounding event occurred, one that got minor press but that may have enormous import. For the first time in perhaps 2 million years, open water existed late in the northern summer in the Arctic seas, opening up the long-sought Northwest Passage from Europe to Asia.[36] Normally the hard winter ice is thick enough to last through the summer melting season. But in 2007 a trend toward greater summer ice melts jumped off the charts. There was palpable fear that one of the dreaded "tipping points" had been reached, and thus there was great interest in the degree of melt that 2008 would bring. As the Northern Hemisphere's summer passed, it was clear that once again an extraordinary ice melt was under way. But at the end of summer the final tally fell just short of the 2007 figure, although it was nevertheless far greater than any year other than 2007.[37] Two years does not constitute a trend, but this situation bears watching. Soon the vaunted Northwest Passage will be a reality every summer. Yet while such a route would be a boon to shipping (and offer an opportunity to exploit mineral and oil deposits beneath the onetime ice cap), this loss of a solid, yearlong ice cap will have radical effects on the temperature feedback mechanism described earlier in this chapter. The loss of Arctic ice will warm the world.

With all of the ice melt in the Arctic, did the level of the sea rise? Happily, no, for the ice that melted was sea ice, already floating in the liquid sea, and its conversion to ice and back to liquid has no net effect on sea level. As an indicator of climate change, however, its lessons and message are stark and clear. A vast melting is under way.

As the ice melt in both Greenland and Antarctica moves deeper into the land mass, away from the coastlines, icebergs will cease to form, which will produce warmer onshore winds that will cause the more interior parts of the ice sheets and mountain glaciers to melt all the faster. Soon two places on the planet will produce what will be by that time one of its precious treasures: fresh water. That water could in theory hydrate Africa, Australia, the American Midwest and Southwest, and China. The problem is that all are either already experiencing drought (China, Africa) or are predicted to suffer droughts in the near, warmed future. But moving the volumes of water on the scale needed to surmount those droughts, and the vast famines that they will have precipitated, will be nigh on impossible.

Polar melting will be rapid, and so much water has to go someplace. Much of it will return to the sea, perhaps in such quantity that the large thermohaline currents, so important to redistributing heat (and to keeping Europe anomalously warm), will be shut down. But in great areas of both Greenland and Antarctica barred from human penetration by their topography, enormous volumes of freshwater will become locked on land, resulting in the formation of enormous, sea-sized freshwater lakes in the middle parts of the two landmasses. In this way, following the last great melting, of 15,000 years ago, were the Great Lakes formed in North America, and Lake Baikal in western Asia. And as the ice melts, the land upon which it so long sat will begin to rise by isostacy; the great weight of ice had compressed the land, and with its removal a glacial rebound will begin.

The melting fronts of the glaciers will be extraordinary, and extraordinarily harsh places that will change nature, making it unforgiving but also ultimately greener. Incessant, strong winds characterize the retreating glacial walls. So strong are these winds that they will create great piles of sand and silt to be carried by the wind, in the form of sediment called loess. But the winds will also carry in seeds, as will birds, so that the drifting soils in front of the glaciers will soon be colonized by pioneering plants. First will come the ferns, hardy in the very poor soil that emerges from under the ice, and then more complex plants. In Greenland we can expect willow, juniper, poplar, and a variety of shrubs to be the first stable species to transform the ancient glacial regime; and soon thereafter successive communities of plants will arrive, depending on location. In the more temperate western part of Greenland, low forests will become dominated by spruce. In the middle, colder parts of Greenland, permafrost and tundra will be the norm for long periods of time. But eventually this will pass as well. A vast new agricultural region will emerge.

The kind of plants eventually colonizing Antarctica is more difficult to predict. The nearest areas that could provide seeds would be New Zealand, Australia, the southern tip of South America, and South Africa. Because each of these has its own endemic plants, the mix in the new Antarctica could be novel.

Who will claim the polar riches of minerals and oil, and the potential riches of agriculture in a starving world? Denmark "owns" all of Greenland, and might become one of the world's wealthier countries if it can maintain its grip on its colony. But more powerful and envious countries in far more dire straits than Denmark may well step in and just take it.

The situation in Antarctica will be more complex, at least legally. The newly emergent continent might be split up into separate territories of those countries nearest, or richest, even though they are not particularly close, such as America, Russia, and by this time, perhaps China and India. The latter two will be in the most precarious position, for the loss of Himalayan glaciers long before the warming of the poles will have plunged both places into drought, perhaps leading to mass human mortality measured not in the millions, but in the hundreds of millions, or even low billions.

Ultimately, Greenland and Antarctica will become ice free—and in so becoming, they will have caused sea level to rise more than 200 feet. With their loss, the final act of human-engineered climate change will play out. What began as an industrial revolution in the 1800s and progressed to an oil economy in the 1900s, only to end as a coal economy in the 2100s and 2200s, will offer us a recipe for potential human extinction.

CHAPTER 6

FLOODING OF COASTAL COUNTRIES AND CITIES

Zwolle, Netherlands, March 2100 CE. Carbon dioxide at 650 ppm.

There had been no sound for two days but the furious howling of the great North Sea storm that was punishing the Low Countries with terrific force. Elsewhere it would be called a typhoon or hurricane; here it would be simply labeled just another storm of a kind increasingly common. This entire region was progressively falling beneath the official sea level, a moving target if ever there was one. Huge areas around the town of Zwolle were now threatened by storm surge that could surmount the extensive sea walls and dikes erected in recent years along the coast. The people of this region of Holland had long since given up worrying about whether the dikes would hold. They simply hoped. Although the huge investment in saving Holland from the sea might not entirely succeed, it had at least lessened mortality from catastrophes produced by storm surge. But that was about to change.

In the late twentieth century, the Dutch had sealed off the entire Zuiderzee with a huge dam that turned it into the large, freshwater Lake Ijssel.[1] Later, in the twenty-first century, they built huge levees to tame the Rhine and Meuse rivers as well, with enormous storm surge gates erected at their mouths. The structures had one goal: to keep the sea at bay. These public works were extremely costly, making taxes in Holland and Belgium the highest in the world. Many aspects of the socialist services expected in northern Europe were fraying as more and more of the gross national product of the Low Countries went into the expensive dikes and coastal defenses—outlays especially burdensome at a time when the continent's population was declining due to low levels of reproduction. But as the water defenses grew more massive and spread across an ever-greater length of the long Dutch coastline, there became more and more dangerous sections, where the sea would not be thwarted by the modern equivalent of a boy's single finger in a dike. The army was now a trained dike-plugging force. The Dutch knew all too well the cost of failure: way back in 1953 a storm surge had smashed through the then-primitive sea defenses in southern Holland and nearly 2,000 people had drowned.[2] That flood, whose high-water mark was still visible on Amsterdam's city hall, drove water levels from 15 to as much as 18 feet above normal. But then again, sea level had been about 5 feet lower at that time, compared to what it was now at the start of the twenty-second century. The old "normal" sea level was now well underwater.

Melting ice from the Alps had always made its way northward across Europe, creating flooding in the spring in Belgium and the Netherlands, both of which had numerous rivers draining into the sea. But in this new age, the floods could come in the winter as well, and the unlucky coincidence the Dutch feared most was the confluence of a hurricane-strength storm from the sea joining a major flood from the rivers. That possibility seemed ever more probable, because global warming had made the altitude allowing snow to become so high that it rained rather than snowed most of the winter. The traditional spring thaw no longer occurred, where snow accumulated all winter in most mountains, and stayed there until the spring temperature increase melted it all, as it did in the Sierras, Cascades, and Rockies. But in a globally warmed world, where temperatures were now so high that rain predominated, even high-mountain winter precipitation was in the form of rain rather than snow. Because of this, the snow level in the Alps became so high that winter precipitation once falling as snow now fell as heavy rain. This in turn caused the multitude of rivers that weaved through the country to reach flood stage more frequently than in the past. The Dutch and Belgians were faced with a dou-

FIGURE 6.1. Holland's coastline.

ble challenge: getting the rivers to course to the sea and not flood the very low landscape, while at the same time keeping the sea offshore.

River floods posed as great a danger as flooding from storm surge. For example, a 1995 river flood in the Low Countries (Netherlands/ Belgium) had forced the evacuation of 200,000 people and millions of animals from endangered areas. Much of the countryside would have drowned under muck and water in the continuous rain if pumping stations hadn't won the struggle to drain water faster into the sea than the rivers churned it over their banks. Since then the Low Countries had fought that two-front war, with all of the disadvantages any such fight brings to a military campaign. Like the long-ago Roman emperor Caligula, who in his craziness declared war on the sea itself (actually its god, Poseidon), the Dutch—and the rest of the coastal countries worldwide—had in essence declared war on the sea. But their mode of defense was static, and like the French Maginot line in World War One, their barriers were outflanked and enveloped by storm surges. It was a long fight, bitterly expensive, rife with casualties, and a holding action increasingly hopeless, as the financial resources of Holland and other coastal countries went to build dikes and flood control that was swallowed and made inoperative by ever-higher surges from ever more violent storms.

The Dutch had no choice. Since the advent of this new, twenty-second century, some 70 percent of the country's economic output had continued to be generated on land that was below sea level.[3] Its production areas were protected by an ancient if complex system of dikes now mated with modern cement barriers to hold back water from the sea, yet letting the rivers pass through peacefully.

Lessons had been learned; not all was bleak. By 2100 there were novel kinds of defenses as well as traditional dikes in use. Rather than dredging sand to maintain beaches, the Dutch now routinely dumped piles of sand offshore to create "sand engines" that would be shifted by the tides and shore up the coast. Marshes were renewed, to break the power of incoming waves. Some Dutch even tried to join rather than beat the enemy: the Het Nieuwe water project was one such effort, a half-mile-wide development near The Hague that included a series of floating apartments designed to rise and fall with the water level; the first phases were actually built in the early twenty-first century.

Out of necessity the Dutch had begun a campaign called "Room for the River," which weakened certain levees to re-create natural floodplains along rivers, including the Rhine, the source of the 1995 flood, and its tributaries—a kind of triage that would lessen the brunt of large floods.[4] But those dikes that could not be allowed to fail, those proximate to large populations, were only as good as their foundations, and any major river flooding that overwhelmed them could now submerge almost half the country in 20 feet of water. If the nationwide network of pumping stations failed, within a week the *entire* country would be covered in 3 feet of water. One irony was that the Dutch had to resort to coal, still the only "cheap" form of energy, to keep the energy-hungry pumps working. Out went the water, but up went the level of atmospheric carbon dioxide. The sea continued to rise, slowly eating away at the proud old country. But what else could the Dutch do? It was not time to abandon their country wholesale—but that day was coming. The Dutch, more than most people, knew it. Already they had lost prized territory.

The casualties of climate change in the twenty-second century were not just individuals but whole towns. Leeuwarden, a once-thriving town in the coastal north, had been abandoned and given up to the sea in 2075 along with a huge expanse of northwestern Holland. The populace had been increasingly devastated by the storms, the final blow coming in 2075 when dikes had failed and 45,000 people had drowned in the raging chaos as the storm surge had overtopped the flood control gates and then battered them down, that short-term failure loosing the sea onto the residents of Holland on a dark, rain-lashed night—much like this particular night twenty-five years later. After that, despite the strong objections of affected landowners with salted fields and marshes where once prosperous houses and markets now stood empty, the exodus began.

The day before today's storm, the radio had warned that it was larger than any of the others that had pounded Zwolle in the past two winter months of the storm season. The authorities ordered an evacuation of the town, unprecedented for this part of Holland. It had always been thought to be safe, being so far to the east of the old shoreline and of the new inland sea that had been northern Holland. But this was a new reality. The roads were indeed filled with the fleeing multitudes. Yet, for a million rationalizations, far more stayed behind. Calls for evacuation, just as in the hurricane-prone regions of the United States, were now routinely ignored. People just rode out the storms, and while there were always casualties, most people survived. The alternative here was to move into ever-shabbier and crowded living quarters hard against the border with Germany, the last high ground left in Holland.

The Dutch had prided themselves on producing dikes able to withstand a once-in-a-thousand-year storm. The storm of March 2, 2100, was nowhere near as strong as the most powerful storm of the millennium could be. But it was powerful enough. Storm surge broke the outer seawall defenses in the southern part of the country, just as it had in 1953. But unlike that event, which had been caused by the rampaging sea alone, this storm came amid a historic high level of discharge of the Rhine and Meuse rivers. In a single night, the water came from two directions, the sea and the flooding rivers. The long banks holding in the rivers became too waterlogged, and they failed. Freshwater met the advancing seawater near Amsterdam, and before people could escape it, water quickly rose 20 feet above the city streets, forcing survivors to move onto rooftops, there to wait bitter days for help that was all too slow in coming. In a single dark night, over 3 million people drowned.

THE GREAT GLOBAL EXODUSES TO COME

Where do you go if your neighborhood is condemned, or otherwise made unlivable? Perhaps it is just your luck to be in the way of that new airport runway that has been planned for years, or you find that your house was formerly a meth lab and is still seeded with toxic fumes, or you discover that five blocks in all directions around your house is a toxic-waste site. If you are a prosperous enough individual in an economically stable nation, you can rebuild your life elsewhere, even at great personal cost. But if you were one of over a million Chinese who were displaced from the site of the 250-square-mile lake produced by the Yangtze River's Three Gorges Dam project, a new reality of housing, neighborhood, employment, and many other changes was evident.[5]

The rising seas will uproot not just individuals but entire regions, devastating economies and lives. Because the future of ice sheets and ice caps determines the fates of most coastal cities, and even whole countries, we need to examine the places that could lose land, food, and wealth. Many cities, regions, and nations could be flooded with devastation, including the Low Countries, as well as so many cities that hug the coastlines of continents, but none will be in more dire straits than Bangladesh. A relatively small and extremely poor country, Bangladesh is dangerously squeezed between regions of far more assertive topography. As can be seen in Figure 6.2, the neighboring Myanmar and India surround the low-lying Bangladesh like a wall.[6]

Bangladesh is a bastardized country, formed from scraps and leftovers of twentieth-century political upheaval and war. There was no Bangladesh before 1971, when it claimed independence from its non-contiguous national partner Pakistan. This twinship was based on a unification of its people by language and religion during the postwar partition of the Indian sub-

FIGURE 6.2. Current topography of Bangladesh, in meters above sea level.

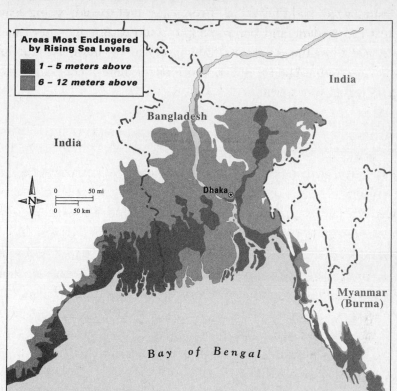

Areas Most Endangered
by Rising Sea Levels

■ 1 – 5 meters above
▨ 6 – 12 meters above

India

Bangladesh

India

0   50 mi
0   50 km

N

Dhaka

Myanmar
(Burma)

Bay of Bengal

continent. It is a place packed with humanity, and a place of low overall elevation. According to Robert Kaplan, virtually every tillable acre of land in Bangladesh is already being farmed—there is no frontier of any kind that could still be exploited for either living space or agriculture.[7] The number of inhabitants in the country is eye-opening: more than 150 million, with the 200-million mark to be surpassed in a decade or two if births continue at their current rate.[8] While the astounding figure of seven children per woman of childbearing age has been cut to only three children per woman, Bangladesh's birthrate ensures that food and living conditions are threatened—and that climate-warming pollution will rise, and rise quickly. All those people have to live somewhere, and while the population rises, the land area will shrink. In fact, it is estimated that an average of about 2,600 individuals reside on each square mile of the country, giving Bangladesh the highest density of human population anywhere in the world.

Bangladesh is not exactly a world ski capital. It has no mountains or even much in the way of hills. It is a vast, flat plain, with land somewhat above sea level only far inland toward its international borders. Because of this, even the smallest sea level rise will have major consequences. It is estimated that even an 8-inch rise in the Bay of Bengal will displace 10 million Bangladeshis. Not all of the land around the bay will be flooded out and underwater of course, but as we discussed earlier, any rise in sea level brings shoreward and inland the dangerous effects of storm surge and wind damage. Typhoons grow in size over warm seas and calm down over land. By removing land at the edge of the country, the small rise in sea level brings the eye of the storms right onto the most populated parts of the country.

Eight inches. That small increase could occur in the next decade; it will inevitably happen well before 2050. Thus the clock is ticking on what Kaplan described as the most likely spot for "one of the greatest humanitarian catastrophes in history."[9] In fact, University of Washington climatologists rather pessimistically, but perhaps all too accurately, have predicted that concerted efforts to stem climate change will not happen until we experience a mass mortality.

The nation's geography portends its doom. Bangladesh straddles the giant delta formed by the convergence of the Ganges, Brahmaputra, and Meghna rivers as they intersect the sea. This geologically young delta is the reason so many humans settled here in the first place, and it is like so many around the world: a region with rich soil, brought into the various deltaic lobes by the many seasonal floods. But unlike some other deltas, this one lies under a tropical sun, and the combination of rich soil, abundant freshwater from the copious rainfall, and tropical heat causes plants to grow quickly and spectacularly, yielding abundant food.

The yearly flooding is a consequence of both runoff from snowmelts in the neighboring Himalayan Mountains and the yearly tropical monsoon season. The floods also cut the country off from the rest of Asia, and this more than anything forged its people into the entity that would demand national independence. Currently, the amount of freshwater is not a problem, but soon it will be. Because much of the river flow through the country comes from the snow and glaciers of the Himalayas, any change in that supply will affect Bangladesh considerably. As with so many other locales that had continental glaciers, the rising temperatures are causing rapid glacial melt. When the Himalayan glaciers are entirely gone—which seems to be their fate—there will no longer be a spring runoff to enrich the delta and supply freshwater to people, animals, and crops alike.

The cutoff of river water will likely come at the same time another villainous effect of rising sea level begins to importantly assert itself: salt infiltration. Salinity might be as dangerous a force of global warming as rising water levels, because salt is already inching inland to underground aquifers of freshwater. The salinization of wells supplying human drinking water, as well as the water so necessary for agriculture, threatens trees and crops, and the problem is here to stay.

All this impending danger is not lost on Bangladeshis. Yet while the country's citizens should be doing everything possible to muscle up protection from the rising sea, the government remains ineffectual, and as a response to economic and civic paralysis, rising Islamic sentiment may again fuel the extremism across the country that was evident in the recent past. Today the nation is more peaceful, but this seems to be a calm layer atop a cauldron set to boil, heated by the nation's grim and persistent poverty. Its people are preoccupied with present-day needs. Routine crises ensure that little if anything is done to master a perilous future.

As cruel as the 8-inch sea level rise will be, it will only be disaster's high-water mark above present-day levels—which we can hope will not occur before 2100—15 percent to 20 percent of Bangladesh's land area will be underwater, with a far higher percentage affected by storm surge and salt infiltration. Some experts have predicted that as little as a 2-foot rise in sea level will actually flood 20 percent of the land area.[10]

Bangladesh's poverty will only exacerbate the problem. That other threatened low country, Holland, has met the rise of sea level with hugely expensive but effective systems of dikes, dams, and other constructed boundaries. Such infrastructure has never been built along coastal Bangladesh, nor is there any plan—or money—to start such construction. Because of this absence of defense, storms and floods have historically taken a huge human toll in Bangladesh. Any dikes that would be constructed in the Bay of Bengal would have to be large. Right now the sea effectively rises over 22 feet during a storm surge—a height that will increase to over 30 feet with even a 3-foot rise in sea level. Any coastal dike would have to be higher than this to protect the populace. Otherwise, when the great typhoons come, we will see a situation like New Orleans after Hurricane Katrina, only greatly magnified.

Even geological forces seem to conspire against Bangladesh. Because of tectonic forces, the crust of the earth underlying all those people is actually subsiding, just as sea level is rising. As a result, sea level is rising two to four times faster than everywhere else.

PREVIOUS SEA LEVEL RISE IN EUROPE'S LOW COUNTRIES

Of all the world's regions, the place taking the threat of sea level rise most seriously is northern Europe, especially in Holland. The northern European coast is built on soft sediments of peat, sand, and clay; it is thus easily eroded. Further, there are no bedrock features, no providential mountain chain separating vast low plains from the sea. Even before an increase in sea level, the well-named Low Countries were in trouble. Now that the seas are rising ever faster, they are in more dire straits. And, not surprisingly, they are perhaps most prepared for what is to come— even if they will eventually lose the battle.

Along much of the northern European coast, long sand bodies composed of dune ridges and high, parabolic sand dunes separate the sea from the flat interior landscape.[11] These shores have long been settled by humans, and within these dunes and at other archaeological sites, a long history of sea level rise and fall has been recorded. These deposits also provide a stark measure of how fast sea level can rise during periods of rapid ice-sheet and continental glacial melt. Geological research into these deposits should give pause to anyone who thinks we have all the time in the world to contend with sea level rise.

The rate of sea level rise in this century is pegged at a measly one-tenth of an inch per year.[12] But careful dating of sediment levels from a variety of northern European coastal sites shows how much faster it could be, under natural conditions, not just thanks to human-engineered carbon dioxide levels. While it has been argued that the maximum rate of sea level rise anytime over the past 10,000 years is about an inch a year— a full order of magnitude higher than that occurring today—the European sites show higher rates. In England and Denmark, rates of rise as high as

2 inches per year occurred over a 600-year period from 8,260 to 7,680 years ago. In Germany, the sea rose at a rate of 1.5 inches per year at that same time. These rates translate to about 16 feet of sea level rise in a single century. Such an increase far eclipses the approximately 3- to 5 feet of rise expected over the next century, and shows that rates could be far higher than currently suggested.

What makes these rates especially frightening is that they occurred during rises in CO_2 that appear to be neither higher in rate nor in absolute value than those we are witnessing today. All this change occurred when CO_2 rose between 100 and 150 ppm over the baseline values.[13] We have seen that increase in the past century alone, and much higher levels seem to be in store soon.

Let's go back to that time, some 8,000 years ago. Imagine living along the seacoast of northern Europe, not a particularly secure or productive place to live in any event. Such a gray, overcast landscape would have featured a short growing season and long, dark winters. Now add to that dynamic a new, strange ascension of the sea.

Would the increase have been noticeable to the locals? Even within a single generation of what surely were relatively short-lived humans, perhaps not. But at the maximum rates there would have been about 10 to 20 inches of sea level rise in ten short years, and about 4 to 8 feet of vertical rise in fifty years—rises that would command anyone's attention. In some areas, the rate of flooding would have been even higher. In Germany, the coastline advanced a staggering 150 miles inland in 1,500 years. This would have affected every civilization that had a coastal presence. No wonder there are legends of great floods.

THE CURRENT VULNERABILITY OF THE NETHERLANDS

As noted in several places in the pages above, of all places in Europe, the Netherlands is most vulnerable to the impending sea level rise.[14] About half the Netherlands is situated below current sea level. This large expanse of countryside is protected by a complicated and widespread series of locks and dikes. Dike building has been a way of life for the Dutch for some centuries now, and they have experienced their share of successes and failures. A great deal is at risk economically—the central part of the

Low Countries (including Belgium) is home to most of the important industrial activities, and no small part of the agricultural areas. In addition, the major portion of the Low Countries' population is settled in this region, in the big cities of Rotterdam, The Hague, and Amsterdam. The response to this potential threat has been to construct an amazing, and amazingly expensive, system of dikes and flood controls. And such effort is not just theoretical. After the 1953 storm overwhelmed coastline defenses in one region and killed more than 1,500 people in a matter of hours, the government vowed to defend the coastline. It is a long coastline to protect—almost 250 miles—and unfortunately for the Dutch, it experiences a relatively high range of tides. For much of the shore, tide change is between 5 and 7 feet, but in summer it is as much as 13 feet. It is the combination of large storms during high tides that poses the greatest current threat, and sea level rise will only exacerbate this union.

Oddly, while the Dutch are clearly worried about the physical effects of current sea level rise and the potential of flooding, as yet there seems to be little discussion about concrete steps to confront a future that promises wholesale change in their land, economic and social dislocation, and social uproar. For instance, even if the land is not flooded, the change to the water table and responses of plant life to the first appearance of salt in what had been fresh groundwater can spell economic ruin, and will spell economic ruin for those who will try to ride things out with the crops and livestock that succeeded so well in the previous century. But even more pressing, yet still uncommented on at a national level, is the response that Holland (and many other countries) must take if and when the ice caps begin melting in earnest in the foreseeable future. Perhaps the present is bad news enough, but planning for the rise is not happening on the scale necessary to manage it in the century to come. A 3-foot rise in sea level will put Holland on the ropes; a 16-foot rise will knock it out, flooding huge areas of the country and displacing millions. Where will they all go? How will they live?

The rising of the ocean as much as 16 feet above present-day levels would be catastrophic the world over. World leaders must plan for these events if their nations are going to handle the coming disasters. Unfortunately, the very nature of politicians and the people they serve mitigates this proactive response to climate change. For instance, four Dutch scientists in a 2005 article posed the question of political response to the very real threat to Holland, and noted this about the political class:

To be a successful politician, it is important that, to the public, one appears successful and is a high achiever. Being perceived as successful depends also on the prevailing political feelings in society, such as about solidarity, nationalism or globalization. Big political decisions must fit into such moods. A second consequence of the working conditions of politicians is that they will not bring bad news [of accelerated sea level rise]. Governments will continue to radiate trust and adequacy.[15]

Rising sea level will be a political problem as it becomes more of a present-day reality. Governments both national and local will not easily think of abandoning any territory and forcing people to migrate. Governments are inherently conservative because they must account for their actions. As a rule, politicians should be more responsive to public opinion than to warnings of scientists.

At least Holland is thinking about the impending rise of its waters—to the extent that a number of scenarios have been formulated that could be used to forestall it. It is useful to summarize these. For Holland's best-case scenario, imagine that the year is 2030—not very long from now. The country's population has risen from 16 million in 2008 to 18 million. Half of the population lives below sea level.[16] Fully half of those 18 million are retired and elderly. Sea level is somewhere between 4 and perhaps even 10 inches higher than now. To deal with the oncoming flooding by storm surge, Holland has completed about 930 miles of "primary flood defense" structures: walls and dikes to hold back the sea. Because of this progress, achieved at high monetary cost, the public is not worried about sea level rise and flooding—it is assumed that the problem has been taken care of and that any possible flooding will be a local nuisance, not a countrywide disaster. Politicians have the same idea. The government's major response is to compensate people for land ruined by flooding.

Jump to 2070. Sea level has risen 3 feet above 2008 levels. Local flooding occurs every winter in the stormy season, but there has been at least one real calamity caused by failure of a dike. In that case, the dike just north of Amsterdam was breached during a major storm, with great loss of life. This tragedy wakes up the public to the danger of rising waters. Much of North Holland is now flooded, and this land cannot be easily reclaimed. Even if the dike is rebuilt and the water pumped out, this largely agricultural region has been ruined because of the salinization of its soil. Holland recognizes finally that this event is a harbinger of its future. At about the same time, Venice floods and is abandoned, further heightening awareness of Holland's plight. The nation's economy stagnates because of both financial losses from the flooding and widespread panic. Land values for higher ground skyrocket, pricing many people out of the market. Lowlanders and highlanders become political groups and divide into separate camps. The government decides on a cost-benefit strategy for future defense, because sea level is still rising, at an accelerating rate, and now threatens the entire economic system. Strategies now include planning the protection of urban centers that occasionally flood. Temporary evacuations to higher ground are mandated. However, the possibility of permanent evacuation of the still dry but low-level lands is discussed but not acted upon.

The southwest and north of Holland are the most threatened in 2070, with even greater loss of land anticipated in its northern region.[17] The government decides to put all its engineering and capital into saving cities at the expense of abandoning agriculture in threatened areas, which enlarge yearly. Some of the loss of gross national product will be compensated by new engineering firms that export the Dutch flood control system to many parts of the world. Because all coastal countries are threatened (yet few, except Bangladesh, have experienced the loss of life and land that Holland has experienced), sooner or later all of them will try to save their coastal cities, infrastructure, and agriculture. Holland realizes this and even now attempts to integrate its economy more firmly into that of the European Union. But no country has yet offered to relocate any of Holland's citizens.

By 2130, Holland's geography will be very different from what it was just a century earlier. Those areas that had been flooded only by storm surge are now permanently underwater. Agriculture is a minor part of the national economy, yet remains an important aspect of political parties.

There has always been a very strong farm lobby making sure that farmers are helped by government—a lobby that will simply grow more strident as farmland disappears. This will lead to widespread food shortages and even famine in certain years, because of the loss of worldwide agricultural lands to flooding coupled with more human mouths to feed globally. Holland's economy begins to lag behind that of the European Union as a whole. Holland appeals to the EU for refugee compensation—which would allow payments to those affected—in essence, European-style welfare to the citizens of a blighted country. The lowlanders become a strong political party, with the votes to delay the emotionally charged issue of permanent abandonment of their home regions, as many who have been displaced still call on the government to push back the sea. But as time goes on, the economy of lowland areas diminishes, as does the regions' political power. One-third of Holland was below sea level at the start of the twenty-first century. This area, coupled with above–sea level areas that now find themselves below the new level of the sea, comprises greater than 50 percent of the country. Ultimately these low areas are officially abandoned to the sea.

OTHER THREATENED REGIONS AND CITIES

Holland is far from alone in facing the danger from land loss (and greater human mortality) due to sea level rise. In fact, thirty-three countries have land below sea level. Many of the areas below sea level can be classified as tectonic depressions, downward flexes of the uppermost crust of the earth due to plate tectonics. Included in these depressions are big cities (New Orleans the most famous of them, having so disastrously flooded even before climate-induced sea level rise) and even a few of the "megacities"—those with more than 10 million inhabitants, such as Bangkok.

The world's ten lowest depressions are the Dead Sea Depression, at 1,355 feet below sea level (BSL), Lake Assal in the Afar region of eastern Africa (509 feet BSL), the Turfan Depression in China (505 feet BSL), the Qattar Depression in Libya (436 feet BSL), the Karagliye Depression (410 BSL) and the Denakil Depression in Ethiopia (394 feet BSL), San Julian's Depression of Argentina (344 feet BSL), Death Valley in California (282 feet BSL), Turkey's Akdzhakaya Depression (266 feet BSL), and the Salton Trough in the United States and Mexico (226 miles of sub–sea level). Some are threatened, some, despite their very low elevation, are not, simply because they are walled off from the sea by high-altitude topography.

And not just regions are threatened. Most endangered of all are coastal cities. As sea levels rise, all will be affected. Some, such as Seattle, Rio, and parts of Los Angeles, will be less affected than cities with low elevation only, saved by the steep topography they enjoy. Many others will not be so lucky. In a fictional anticipation of one potential reality, near the end of the 1999 Steven Spielberg movie *Artificial Intelligence*, the young robot protagonist finds himself confronted by a drowned Manhattan.[18] With waves washing through windows clearly many stories above what was once sea level, this might be the most accurate vision of most of the world's coastal cities if all the ice sheets vanish. It will not just be agricultural fields that will bear the brunt of the rising sea. Yet it will take centuries of rising waters for this urban drowning to occur, and undoubtedly the cities will fight back through various expensive and not-always-futile engineering schemes, which will be discussed in the final chapter of this book.

Of all of humanity's constructions, none is as complicated as a modern city, and none is more in danger from the flooding to come. They are massive enterprises, held together by complex webs of water, energy, sanitation, roads and other transportation infrastructure, political rules and ties, laws, planning, economic interests, and so on. The modern city is at the zenith of human creation. They are like some sprawling kind of life unto themselves. In a way they are memes, that quasi-evolutionary entity spawned by genes that can undergo their own evolution, and certain cities have done just that—evolved. But like a species, they are also susceptible to the opposite of evolution: extinction. Cities are born and ultimately die, and never in history has a greater number of cities been threatened with extinction on such a large scale.

All of the coastal cities can die by drowning. If we do not act, none will be spared, even those that climb up hills onto steeper slopes, such as San Francisco, Rio de Janeiro, and Vancouver. The drowning will cover such important organs needed for a city's survival that they will die.

TABLE 6.1. Elevations of various Manhattan streets.[19]

Geographic Place	Feet Above Sea Level	Event Needed For Flooding
Central Park Reservoir	100	Greenland ice melting, and better part of West Antarctic Ice Sheet
8th Ave and 53rd St.	60	Greenland ice melting, and better part of West Antarctic Ice Sheet
Washington Square Park	20	Greenland melting
3rd Ave & East 14th St	28	Greenland melting
East Houston St. and Avenue D	3	Current CO_2
Bleecker St. and the Bowery	35	Greenland melting, and first part of Antarctic ice sheets
Broadway and East 23rd St	33	Greenland melting, and first part of Antarctic ice sheets

Coastal cities face three stark choices. First, they can do little or nothing and become inundated. Second, they can protect themselves by flood control measures and great dikes—for as long as they can. Third, they can relocate. But relocation does not just mean that cities can pull up their roots and replant themselves some distance uphill, plopping down in some semblance of what was there, even if they somehow managed to take some of their iconic buildings and all their park statuary with them. These relocated metropolises would be entirely new cities. More likely, the cities would simply be abandoned, not somehow stuffed into the guest bedroom of another city, like relatives who lost their house in a flood.

The fate of the coastal cities will be decided by their elevations. Cities perched on hills beside the sea will save some portion of themselves. Those at sea level or nearly so will be lost. Some may be lost partly, or in increments, as would be the case in New York. In Table 6.1, some locations in Manhattan are listed as mere feet above sea level.

There are many examples of what is to come, but perhaps the most relevant are those cities already being flooded because of environmental factors other than sea level rise. The most instructive are Venice and New Orleans.

VENICE, DROWNING

The great maritime city of the Italian Renaissance was built on soft sediment, and its heavy buildings just contribute more mass to an already subsiding land area, enhancing the rate at which it sinks into the mire. Venice is particularly susceptible to climate change. In 1900, St. Mark's Square flooded about ten times a year; now it is about sixty times a year. The water level in the city is permanently too high nowadays—10 inches above the mean water-level reference point established in 1897—and is already eroding the brickwork of the buildings.[20] Venice is not just any city; it is one of the most beautiful works of humanity, a cultural monument that needs protection as much as any United Nations World Heritage Site. The realization that something must be done to save Venice was most urgently recognized in November 1966, after the city underwent a devastating flood—the kind that will soon be the norm, rather than occurring once a decade, in Venice. The UN Educational, Scientific, and Cultural Organization (UNESCO) and some international private committees began campaigns to raise funds to help protect Venice. Successive Italian governments publicized this problem, essentially thrusting it into the international arena.[21] This was a novel approach: asking the world to spend money on a city that was and is highly endangered by rising sea level. While Italy declared the city's situation to be "a problem of essential national interest" in a 1973 Special Law for Venice, its governments were so unstable and transitory, and its regional squabbling so intense, little was done.

Also in 1973, UNESCO published a manifesto about Venice declaring that "the participation of the international community is a moral obligation. If Venice really represents, in the eyes of many men and women, a vital common asset, then they must share the burden of its preservation. In future years, when other cities come under siege, a summons to such a moral imperative may well not have the same punch." Some aid to Venice was forthcoming. But much of the attention lavished on the "Venice problem" was driven less by a rational response to the objective conditions of flood damage than by international moral obligation.

There certainly was no argument made that the world needed Venice on economic grounds or, for that matter, that Italy needed Venice for anything more than the tourist money it brings in. Venice is not a vital shipping port, or a manufacturer of rare goods needed by the rest of the world or by Italy. Few people live there. It is a beautiful museum, not a living city. By 2006, some were finally wondering whether, indeed, Venice *should* be saved. In the *Times* of London, under the headline "If You Love Venice, Let Her Die," Rachel Campbell-Johnston noted the increasing frequency of flooding in Venice recorded over the past hundred years.[22] Her proposal that we should consign the city to the sea expresses some people's sentiment that humans should not tamper with natural phenomena such as subsidence. This is nonsense, since it was our tampering that got us into this literal hot water.

There is a crucial difference between the current efforts to save Venice and those proposed in the 1960s. The sea did not cause the great 1966 flood. Instead, the flood was the result of high water in the several rivers that drain out of the nearby Alps and converge near Venice. Flood control in the first days of action was oriented toward taming these rivers, to the point of diverting them upstream if necessary. To that end, engineers raised the city's lower areas. But as the twentieth century came to a close, scientists identified the real culprit behind the flooding: sea level rise directly related to global warming. Then engineers designed a series of seawalls to push back the city's potential killer, the Adriatic Sea. This barrier system is due to be completed in 2011.[23]

However, the seawalls came a little late. In 2003, for the first time, a British government official, the influential Sir David King, argued that the barriers would be irrelevant without an international agreement to curb the rise in carbon dioxide emissions: "I am saying in the face of sea level rises already happening, for example, the barrier that is to be built and is currently underway would be insufficient through this century to protect Venice."[24]

Construction work began on protecting Venice from high-level flooding using mobile barriers, which will close the three entrances to the

Venetian lagoon from the sea during high tides.[25] However, the fate of Venice and other coastal cities will be decided by politics. In rejecting the Kyoto climate change treaty, the United States sent a very hostile message to the endangered places on the earth's coastlines. So Italy took the practical route, as Holland had. It decided to wall off the city from the threatening Adriatic.

This work continues on the mobile barriers (estimated cost around 4.3 billion euros), to be raised when abnormally high tides are predicted. But just how long these barriers will be effective depends on the degree of sea level rise. The Consorzio Venezia Nuova, which is building the barriers, has factored a rise of 10 to 24 inches this century into its calculations and believes the barriers have a hundred-year technical life. Yet we now know that this is a woeful underestimate, and in fact the level of the sea might be 10 to 24 inches higher in as little as forty years. Moreover, the fact that the seawalls are mobile rather than stationary indicates that they are intended as defenses against storm surge, not the inevitable flood.

Venice will drown. There is no saving her. What remains to be done is deciding what cultural artifacts to salvage from the city and what to leave behind, and how to uproot and resettle an entire city population without bankrupting Italy's already-fragile economy and political structure.

NEW ORLEANS AND THE NEXT BIG STORM

The city ravaged by Katrina is also living on borrowed time. With so much of its area below sea level, New Orleans was a disaster waiting to happen. It is true that since the 2005 storm, new levees have been erected, built to withstand a Category 4 hurricane—less than that of Katrina, curiously enough. But as is the case in Venice, the decision makers and engineers have built barriers that can withstand storm surge in a major storm. Willfully or not, they ignore that rising sea level will become ever more of a threat to New Orleans, not only from floods but from higher and more penetrating storm surge.

The lowlands between New Orleans and the sea have always been the first line of defense against giant storms battering the city. These vast freshwater swamps and outer salt marshes work to reduce the energy of an oncoming storm (hurricanes always lose energy when coming onto land). Yet

FIGURE 6.3. Areas of the Gulf Coast, Florida, and the lower East Coast that are threatened by even a modest sea level rise.[27]

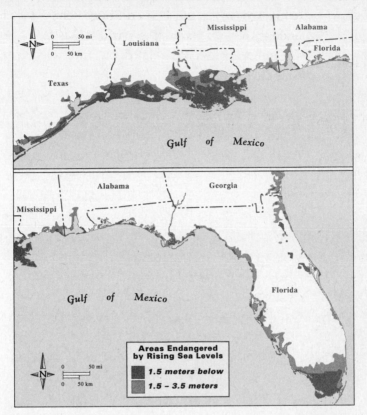

the slow seepage of salt into the freshwater marshes will first kill vegetation there, leading to loss of bank stability; ultimately all of the lowland in front of the city will be under water. The first large storm will finish what Katrina began, emptying the city of its economically mobile citizens.[26] If by some chance there is not a bull's-eye or even a nearby hit by a hurricane in a world where there are ever more of them, the rising sea itself will do the job. The entire Louisiana coastline (as well as certain portions of the Gulf Coast and Florida) is definitely endangered, as seen in Figure 6.3.

THE BAY AREA AND BEYOND

San Francisco Bay is the largest estuary on the West Coast of the North and South American continents. At the time of the 1849 Gold Rush, the open waters and bordering wetlands of the bay covered 787 square miles. But the bay was shallow and is more so now; two-thirds of it is less than 12 feet deep. The Bay Area will be a test case of cities' responses to sea level rise. Those threatened (the largest being San Francisco, followed by Oakland and San Jose, but with many smaller municipalities equally endangered) will be watching to see whether response is united both in plan and money. San Francisco is far wealthier on a per capita income basis (and hence in tax revenue) than is Oakland.[28] Will response come equally to both?

With post–Gold Rush development of San Francisco and nearby cities, the bay began to shrink. Siltation from the shorelines and many rivers flowing into the bay filled it, while shallow tidal areas were diked off from the open bay to create farmland and salt ponds. Cities used the bay as garbage dumps, and numerous land reclamation operations were undertaken to create dry real estate where bay waters once flowed.

By the middle of the twentieth century, the bay's open waters had been reduced to 548 square miles and nearly a third of the bay—239 square miles—was gone.[29] In 1959, the U.S. Army Corps of Engineers published a report concluding that it was technologically and economically feasible to reclaim an additional 325 square miles—60 percent of the remaining bay—by 2020. The local populace rejected the notion that the bay should be allowed to become little more than a wide river.

No longer will the bay grow smaller. It is already enlarging and will rise even higher over time. Bay Area cities, progressive by nature, have made plans to manage the rising waters. At least in this case, unlike in Venice and many other cities, the planning commission acknowledges how unpleasant a job it is to plan for climate change; such commissions like to plan for more development, not less. They also made the brave call that triage will have to be effected; the Bay Area will have to decide what to save and what to surrender. Unfortunately, as in almost everywhere else, planning seems to be for a 3-foot sea level rise. No one in San Francisco or anywhere else seems to understand that sea level will keep rising even after that momentous strandline has been reached.

The Bay Area planning commission made the following statement, which should be a template for all coastal cities:

The first step in preparing this plan should be to determine more precisely which shoreline areas are vulnerable to flooding and storm surge from sea level rise. With this information in hand, the value of both built and natural resources that will be impacted can be determined. Next, the flood-prone areas that are already occupied by development that is too valuable not to protect should be identified. Using this information, a regional flood protection strategy can be prepared to describe all the dikes, levees and other protective devices that need to be built.

The next areas that should be identified are: flood-prone areas where it may be more cost-effective to remove existing development than to protect low economic value structures; low-lying areas that are planned for development that has not yet been built.[30]

The plan set forth these goals in 2008:

Within four years, BCDC [the Bay Area planning commission] should be required to prepare an adaptation plan that includes a strategy for adapting to sea level rise in San Francisco Bay and the Suisun Marsh over the next 50 years. The plan should take full advantage of ecosystem-based management principles to ensure that future development, shoreline retreat, flood protection and wetland enhancement strategies will be coordinated to achieve a vibrant, healthy Bay that co-exists with sustainable communities around the Bay. The plan should determine the measures needed to adapt to projected sea level rise by identifying: (a) The most significant structural, environmental, aesthetic, social, cultural and historic resources that must be protected from inundation; (b) Those areas that are inappropriate for protection from inundation; (c) Those areas that are most suitable for wetland restoration, habitat enhancement and other opportunities that would enhance the biological productivity of the Bay; (d) Undeveloped uplands that are suitable for marsh migration; and (e) Strategies and techniques that will make future conservation and development projects more resilient to climate change.

The BCDC's goals are all to the good but fail to address a central salient point. The planning commission says it is planning for fifty years. But its members never tell us the magnitude of sea level rise the plan anticipates. This is a deadly mistake.

In Miami, New York, Venice, New Orleans, San Francisco, and Rio de Janeiro—among the myriad other coastal cities—the time will come for neighborhood triage: which to fight for, which to give up. Eventually there will be de facto worldwide triage. Ultimately, if sea level rises much past 5 feet, expensive infrastructure will be threatened. The first major blows to humanity from sea level rise will be economic ones. For example, the airport runways in San Francisco; Honolulu; Sydney, Australia; and many other places are built on fill right in the sea or in estuaries of the towns themselves. These runways will be the first to go. Since my city of Seattle just spent over a billion dollars simply to add a new runway at the Seattle-Tacoma airport, a sum that was essentially only for paving, what is a major airport worth if it must be built from scratch? The last major airport to be built on a bare chunk of land was in Denver, and the cost of that facility was almost $5 billion in 1996 dollars, not 2010 dollars—when the dollar was worth a lot more than now.[31] It will prove most cost-effective to abandon the coastal cities rather than to try to protect them from flooding; and planning should be under way to find suitable, undeveloped upland areas. Making these choices will be difficult, particularly when the urban areas contain significant environmental, aesthetic, social, cultural, or historic resources or where their removal would raise environmental justice issues. For example, will all the hazardous-waste sites be removed? Will existing laws allow relocation at all, since siting a whole new city would require a large number of environmental impact studies? Will these be passed or waived? But, in fact, the inexorable rise in sea level will eventually remove all options but one: abandoning the coasts and their cities, large and small.

THE ECONOMICS OF SEA LEVEL RISE

The catastrophic climate changes in coastal areas will reverberate beyond their borders to create enormous economic consequences. The Great

Depression and the world economic plunges of the early twenty-first century will be nothing compared to slow-motion calamity caused by warming skies and creeping seas. Vast sums of money will have to be spent on the consequences of climate change. Yet if there are losers, so too will there be winners. And so too will the balance of global power—military, economic, and political—shift like the sands of the many newly created deserts of our rapidly warming world.

As we have seen, the earth has been warming for thousands of years since the last retreat of the glaciers, and clearly that warming helped humans create and then propagate agricultural and technological civilization. The problem is that the world's current economic and political landscape was largely put in place in medieval times, while the new warming will enforce new rules on the playing fields of today's civilizations. In the past, small climate changes affected local agriculture, transportation, and the kinds of products people produced. Large-scale climate change, however, like that of the Little Ice Ages nearly a millennium ago, brought down whole civilizations, such as the Scandinavian colonies on Greenland. But climate change goes both ways, colder or hotter. A rapid warming in medieval times, from 1000 to 1400, was instrumental in the political rise of France, England, and Spain, as increased agricultural productivity accompanying the warmth allowed these countries to worry less about feeding their masses and more about producing the engines of war and colonial conquest that ultimately gave all three vast wealth and power. Surely the new world we are now entering will have its own enormous effects on civilization.[32]

Our current civilization is now largely based on free trade through globalization, and that, coupled with climate change, will undoubtedly create economic chaos at some level, either locally or globally. Such chaos may bring about a vast redistribution of the flow of wealth.

The intersection of rising sea level, temperatures, and population will surely affect land prices. In areas newly covered or disastrously damaged by the rising sea (such as by higher levels of storm surge and salinization), real estate values for what were once some of the most expensive properties on Earth will fall to zero. Imagine the economic effect when all the wonderful waterfront houses are lost. At first insurance will try to keep up, but as the losses multiply, it is reasonable to infer that eventually the major insurance companies will either fold in bankruptcy or not pay any more claims on coastal real estate, no matter how valuable it was before sea level rise. (Hear that, Hamptons? High water is coming.) Non-urban residential real estate will be profoundly affected too. The Hamptons, Malibu, the Outer Banks of North Carolina, the coastal islands in the Mediterranean such as Majorca and Sicily—at least the edges of all will either disappear or be sown with sea salt. Add up the price of all the world's waterfront houses within 7 vertical feet of current sea level and you get a sense of the monetary loss. Second, if we assume that global warming affects the entire globe, including the tropics (even if dispro-portionately, with the higher latitudes becoming relatively more warmed than the tropics), we should see many of the currently hot, crowded, and poor countries become even less prosperous.[33] Thus there are two great categories of losers: land along seacoasts or facing the sea at low sea level, and the current hot countries.

Among the winners will be locales that today are too cold to be de-sirable for year-round dwelling. Through geographic accident, most such places are in the Northern Hemisphere. The biggest victors will be Canada, Alaska, Greenland, Russia, and Scandinavia, and in the South-ern Hemisphere, Argentina most of all. Perhaps future world power will not be relocated to the countries in the Southern Hemisphere, as is of-ten predicted, but will stay concentrated, if redistributed, in the North-ern Hemisphere.

Here we need to understand how the new warming will be distributed over a daily cycle. If the new climate results in scorching-hot noontimes, with most temperature increase limited to midday, there might be prob-lems for crops even in the newly warmed and fecund places. But studies suggest that much of the new warming will be concentrated at night, rais-ing nighttime temperatures, resulting in earlier springs, and delaying the annual transition from fall to winter. These kinds of conditions will be a boon to agriculture. Farmland in the upper Northern Hemisphere should greatly rise in value.[34] Furthermore, places currently covered by ice sheets or permafrost will lose their icy cover, beneath which will be vast new mineral resources. Finally, all that melting ice has to go somewhere; it will produce many new sources of freshwater, from newly established lakes to streams and rivers.

Food, minerals, and water will translate to power and influence in the world. We can expect both Russia and Canada to become greater world powers. A rich new Russia, with the military muscle to block its borders from populations displaced by rising sea level and the ability to exploit newly emerging crop, mineral, and water resources in neighboring countries, is a frightening prospect. We already see how the Russia of the twenty-first century politically squeezes NATO countries on account of its natural gas production and sales to Europe. With the new world climate, that influence—and bullying—can only increase.

Also frightening is the possibility that the current poor countries, most in the hot latitudes, become even poorer, and even more packed with humanity. They will be fiercely envious of those countries still doing well agriculturally and economically. Imagine Brazil losing its current economic vitality and slumping back to a state of economic chaos and food shortages. South America would spiral into economic and social destabilization.

A reshuffling of resources could also result in great land grabs. We have already seen how Greenland and Antarctica could become battlefields. Such newly emergent and ice-free land will not stay vacant for long. Treaties will be swept away in the land rush made by the rich countries, or even in poor countries with nuclear arms.

Canada could face harsh repercussions. Over the past few decades the Canadian government has ceded most of its Arctic territory back to the original inhabitants, the Inuit and other Native American peoples.[35] Currently these people are peaceful, living as they do on cold hunks of real estate that the European civilizations (including the warmer parts of Canada) would never want. That will change.

Ultimately, what might be the real economic scorecard? The United Kingdom has made one such estimate, concluding that its gross national product will decline by 20 percent within the next century.[36] No estimates are available for the larger countries. For all nations there will be challenges. But there will be opportunities as well. Coastal cities could remake themselves, even if some would have to pull up stakes entirely and head toward higher ground or merge their people, economy, or culture with those of another locality. But just like people who are flooded out by a once-a-millennium storm on a floodplain deemed "safe," the cities will make their next moves cautiously. They will certainly want to know where and when the rise in sea level will happen before their tall buildings emerge from the graveyard of the sea, tombstones of human folly.

CHAPTER 7

EXTINCTION?

September 2045 CE. Osprey Reef, off the coast of Australia. Carbon dioxide at 450 ppm.

The dive boat *Undersea Explorer* approached the vast seamount from the west, having spent the night churning into the trade winds (and thereby making life miserable for all aboard except those with the hardiest stomachs). It was a blessing for everyone to hear the anchor play out, causing almost all motion to disappear as the 75-foot boat came to rest at its protected, back-reef anchorage. The vast reef, none of it above water, now stretched out before them.

This team would be the last tourist divers to come out here, ostensibly because the Australian government had banned the polluting dive boats. But more than that, the tourist-divers no longer wanted to come. There was no longer any charismatic or photogenic life to see here amid the sad death.

Osprey Reef is a seamount rising from the Western Pacific Ocean, an area also called the Coral Sea, and for good reason—at least when it was named in a previous century.[1] As a seamount, Osprey reaches up to the surface like some huge calcareous tube, 15 miles across. Its great central lagoon is filled with giant coral heads that the Australians, owners of this rock, call bommies. The outside of the reef is like a sheer wall, vertical in places, bristly with a great deal of life extending out into the open sea. Not so long before, this, like the many other seamounts dotting the deep oceans of the world, was an oasis of life amid the near desert of the open ocean. But the great volume of large fish living near these undersea mountains had long attracted commercial fishermen, and since most of the seamounts were outside the territorial limits of coastal nations, there was no limit or limitations on fishing, and consequently most of the seamounts had been fished out. But the Osprey Reef seamount had encountered another fate—coral bleaching, and the replacement of its once-abundant animal life, from swimmers to sessile sitters, with gardens of bacteria.

The walls of any reef are their own time machine. If you venture downward from the warm surface waters, the increasing depths of reef wall get ever older in age, with each level older than the one above it. Each level of the reef once laid at the surface, in the shallows where limestone-forming organisms can most quickly build their massive and lacy skeletons in the warm, calcium-rich waters. But the coral creatures are doomed by their own success: the weight of their massive and dense stony skeletons, increasing daily through new growth, causes the reef to sink slightly each year, finally lowering the light-needing creatures—which, oddly enough, include the corals—into the gloom of the depths.

The fastest growers, these same stony corals, can build their skeletons thanks to light, acting on symbiotic plants within the coral flesh called zooxanthellae. These microscopic plants invade the corals early in their growth process, proliferating in the corals' flesh. It is a sweet deal for coral and zooxanthellae alike—the coral develops its characteristic skeleton because of the chemical activities of the zooxanthellae, while the zooxanthellae enjoy an environment rich in carbon dioxide, light, and nutrients that are the waste products of the carnivorous coral polyps. So abundant are the zooxanthellae in the coral flesh that they give the coral its characteristic color, its shade depending on which community of the tiny plants lives within. But to be colorful the tiny plants have to be alive. Such was no longer the case on Osprey Reef, and it was this ghostly appearance of the reef rather than the expected cornucopia of color that met the captain's curious yet sad gaze, now that the engines had quieted, and his part of this voyage was finished until the return to Port Douglas on the Australian mainland some days hence.

The reef had been a diving destination since the latter part of the twentieth century. In fact, some had considered it the best diving locale on Earth because of the warmth and clarity of the water and the sheer underwater cliffs, but mostly for the richness of life here, from the schools of great pelagic tuna and shark to the clouds of colorful reef fish to the carnival of corals that created a kaleidoscope of light and motion. Unlike most of the other seamounts, Osprey Reef was close enough to a continent that it was claimed as sovereign territory—in this case, of Australia. The Australians had done a superb job of keeping the reef pristine for a century to this point. Fishing had been prohibited for decades, so few boats came out. Other than the anchoring buoy, there was no sign here that humans existed at all: no bottles or cans on the bottom, or plastic bags snagged on the shallowest corals, or even the anchor drag marks that characterized so many other reefs. For this reason alone, many people who visited Osprey Reef considered it immune from the smudge of the human fingerprint. But the skipper knew differently. He steeled himself for what he knew would be his last dive on this once-beloved reef.

The familiar gear was like a second skin to the skipper, and he did not bother putting on a thin wetsuit. Once upon a time, the extra insulation was a good idea for anyone venturing below the thermocline at about 100 feet, where cold, nutrient-rich water upwelled and met the warm surface water. But there was no need for the wetsuit now—the upwelling had ceased a decade ago. Now all was warm to the maximum depths that compressed-air scuba could take a diver—and beyond that, the greater depths that mixed-gas rigs allowed. But the skipper did not intend to dive deep today. This was a goodbye, a chance to mourn the passing of an old friend. The few divers with him were old hands of Osprey. They all knew the situation not just throughout these outer reefs but within the entire Great Barrier Reef.

The skipper splashed in off the wide transom without a buddy, popped up, gained his bearings, and headed toward the wall. He knew the dive master could handle the boat's paying clientele, and anyway he preferred going alone. In the crystal clearness, the stony rampart stood stark before him, the depths it sprang from falling away into the velvet black of the deep. As he had on other recent forays here, he searched that gloom for the patrolling white tips, whalers, or even the elusive threshers, but no sharks were to be seen. Nor were jacks or Spanish mackerel, or any other kind of fish, visible. With strong kicks he approached the reef wall. It was colorful, like a heavy piece of furniture painted in many shades of garish green. Here there were fish, of one or two species, busy grazing on the forest of photosynthetic bacteria that covered most of the reef. Here and there longer green algae sprouted billy goat–like like beards that swayed in the swell. But mainly the old diver saw green, like green snot, a gigantic lugee spit on this reef by humankind.

The last time corals had been sighted here was in 2035. That was the date when Osprey Reef's surface temperature had risen above 90 degrees Fahrenheit, and soon that warmth began extending downward year by year, taking out the deeper, fan-shaped gorgonian corals and their near relatives, the lacy alcyonarians, looking like a colony of whips. Sometimes great masses of equally warm water welled up from the dark deep, to swirl and mix with the salt-rich and highly acidic surface water. Those times were unseen until much later—for the warmth, if it exceeded about 90 degrees Fahrenheit, was like a poison to the reefs, leaving behind their figurative bleached bones, in their case the calcareous skeletons, now bereft of flesh.

The entire reef had exchanged tenants; its once-numerous corals had been replaced by microbes that thrived in hot, acidic water. This was no longer a coral reef in the Coral Sea. It was a bacterial reef in the Bacteria Sea.

September 2045 CE. Devonian Canning Barrier Reef. Carbon dioxide at 450 ppm.

Another bacterial reef also received a visit on this winter day in Australia. Across the continent from the Great Barrier Reef, another barrier reef was exposed. But this one lay on land, not in the ocean, and was 366 million years older than the new microbial scum that had smothered the already bleaching corals of Osprey Reef. This fossil reef had never received any official name other than those of its components: peculiar names like Horse Springs, the Oscar Range, Fossil Downs, Guppy Hills, and the most famous of all, Winjana. Now it was variously referred to as the Devonian Canning Barrier Reef, or some perturbation of that unwieldy moniker, a name reflecting its time of origin, and current locale in Australia. It now loomed well above the ocean and along 100 miles of Western Australian outback in a region called the Canning Basin. There, a small plane sputtered and coughed as it bounced to a stop in the tussocked field. The pilot and his three passengers jumped out and secured the plane before trotting toward a Land Rover, dodging clumps of the deadly, needle-like spinnifex grass, the curse of Western Australia. The mood of the men was exuberant—the profligate expense of a private airplane and the ground vehicle ride were novel enough to be heady stuff for them. As members of the Geological Survey of Western Australia, they still had the governmental authority to requisition expensive truck gas, and even more expensive airplane gas, even though lack of resources kept them mostly deskbound. Now it was fieldwork time—the hard and good work of geological study, the brutal heat of the day followed by the chill of the desert air and the welcome swags that made up all Australians' sleeping gear. And the beers before bed, of course.

The flight had been wonderful. After some four hours of fast driving from the Indian Ocean coastal town of Broome across Western Australia on a now deserted and decaying road, they had arrived at an airstrip near the aboriginal town of Fitzroy Crossing. Their flight path had taken them 100 miles northwest along the edge of a prominent ridge of black rock. Looming several hundred feet above the grassy plain, the black rock ran its course for hundreds of miles, its half-mile width undulating like a sine wave. In the early twenty-first century a photographer had taken a picture of this long, snakelike ridge of limestone and then Photoshopped in blue water on either side, turning a picture of black rock on yellow grass into a fantasy version of the real reef lining the eastern coast of tropical Australia—the long, curving outer reef of the Great Barrier Reef. If someone took a picture of that reef and removed the Pacific Ocean, it would be a twin to what once existed here in Western Australia. But the Devonian Canning Barrier Reef, rising between 300 and 500 feet above the trees in front of the assembled pilot and geologists, was 380 million to 370 million years old—the best-preserved example of a fossil barrier reef on Earth.[2] It also held a hard truth: that past global warming had killed a lush and living reef not so different from Australia's greatest natural treasure, and in this it was an ancient double of the Great Barrier Reef in general, and Osprey Reef in particular. Just as was happening off the current coast of Australia, the vast colonial enterprise dominated by animals had been replaced by the form-fitting, smothering remains of a far simpler brand of life, one that could have replaced the animals only under very specific conditions. These conditions, it seemed, were now reappearing on Earth for the first time in more than 200 million years.

It was as if some nasty deity who detested coral reefs and their millions of species of reef dwellers, builders, visitors, and parasites had suddenly drained away a Devonian Period ocean that was then the largest body of water on Earth. Its fish, nautiloids, corals, and shellfish, and the millions of marine species that created or lived around it—the coral skeletons on the deep-water side, and warmer lagoon creatures on the shallower landward side—suddenly had the water emptied from their vast oceanic bathtub, all of it gushing out onto the dusty plain. The sea level dropped quickly and precipitously, leaving the giant reefal edifice to drip dry in the torrid heat. But even more surprising, not only was this reef now bereft of its surrounding water but in the 360 million years since that Devonian ocean had receded from the land for the last time, it had undergone no violent geological change. None of the rock-smashing, animal-compressing, limestone-altering, continent-shifting uproar that is common over geological time happened here. Somehow this part of Australia got a free pass on all the usual torture common to continents and oceans over hundreds of millions of years and untold geological vicissitudes. This ancient reef still undulates across the hot, dry Australian outback, its fossils, bedding, structures, and ghosts utterly unchanged from the time when that unkind god had pulled the plug.

The plane had delivered the geologists to the entrance of the nearly deserted Winjana Gorge National Park, which had been popular when people still caravanned from place to place, enjoying the freedom of the great road. Nowhere had this wanderlust been more enjoyed than in Australia. Its entire populace, it seemed, had gone walkabout at least once a year, making their way to their great parks, beaches, deserts, dry and wet lands, and outback to put up a tent, rig up some camp chairs, in some cases erect television disk antennae, and sit with friends and family for days to talk over cans of Australian beer or bottles of the delicious Australian wines made during the first half of the twenty-first century along the east, south, and west coasts of the continent. The skyrocketing price of a liter of gasoline, and the increasing recognition that driving cars was driving sea level rise, had changed all of that. The Winjana park, like so many built in the time of cheap gas and cheap cars, was peopled now only with the aboriginal community that had prospered here centuries earlier. In fact, they were virtually the only humans living in the vastness of Western Australia, because they did not demand the air conditioning that had become so prohibitively expensive to run that only the elite and the crazy still used it.

The geologists had come to this vacant, desiccated territory to field-test some fancy new gadgets. Traditionally their high-tech machines, the devices that determined chemical composition or the age of rocks, were housed in clean, air-conditioned laboratories, far from the rocks. Yet so many scientific questions needed quick answers: How old was it? Where was it formed and at what temperature? What was its chemical makeup, and that of the seawater or air in which it formed? And if the rock in question was formed by living matter in some way—either through skeletal formation within a body, or through another action of life—could scientists detect the kind of life that had made it—was it from animal, plant, or microbe, for instance? The scientists on this mission had a machine they hoped could tell them much about why the Canning Coral Reef became the Canning Bacterial Reef.

The geologists beheld the enormous rocky wall stretching north and south far into the distance. White cockatoos and pink galahs screamed as they sliced through the morning air in their bickering flocks. The day's flies were also now about, looking for any sweat on the humans. The expedition needed to make its way into the heart of the great fossil reef towering over them. They had to venture into the so-called Classic Face of the Winjana Gorge, a wall of rock named by its first discoverer and still most important student, Dr. Phil Playford, a man dead and buried for more than two decades now.[3] They entered the fossil reef through a tunnel cut long ago by the nearby river through the enormous reef wall, and as they emerged into the sunlight on the other side of the gigantic rocky wall, they found themselves in a canyon of white, brown, and water-stained black rock. Long ago this massive, gently curved ridge of rock had been bi-sected by a river, producing the canyon they found themselves in. And like the Grand Canyon of North America, the river's dissection had cut down, layer by layer, ever deeper in time. Like some monstrous layer cake be-ing sliced through by a giant surgeon, the whole history of the reef and its life lay exposed for anyone clever enough to read its history, written in crystals of limestone, skeletons of the dead, and constituents of the un-told sandy grains imbedded in the ancient reefal landscape.

The same slow, meandering river that had carved this canyon still wound across a quarter-mile-wide valley floor formed by the giant cliffs. The scene resembled a cheesy old fantasy movie of the mid-twentieth century, where the actors find themselves in some lost world populated by either stop-motion clay models or humans in really uncomfortable di-nosaur suits. And indeed, this place had its bevy of fantastical creatures, but they were neither clay models nor rubber suits. Instead, the river was packed with the black, saw-toothed backs of hundreds of crocodiles. None of the geologists paid any heed as they walked over and by the slumber-ing crocs. The biggest was only about 6 feet long, and skinny. These were the docile "freshies" of Western Australia, smallish fish eaters without the propensity to menace humans. They were distant cousins of the other kind of crocodile, the "salties" or saltwater crocs of the tropical ocean coastlines of northern Australia. The salties were twice the size of the freshies, and in the twenty-first century they had been migrating ever farther southward toward the traditionally cooler regions in Australia as the country grew hotter, decade by decade. Now it was not just Broome, Darwin, and Cairns that were benighted by the biggest and orneriest crocodiles on the planet—on Australia's east coast they were appearing south of Brisbane, on a march toward Sydney, while along the west coast the first of them had been found contentedly eating fishermen and their dogs just north of Perth.

A brisk half-hour walk brought the small team to the Classic Face, the largest of the river valley walls. It was a vertical face of layered lime-stone about 400 feet thick. The lower half of this white rock was com-posed of untold broken fragments and larger hunks of archaic corals and the heavy to delicate (depending on the species) calcareous sponges known as stromotoporoids. The colonies of these two animal types had grown one atop another in stony profusion amid the warm, oxygenated water that is characteristic of all coral reefs, irrespective of age. The warm water, rich in calcium and food, was the perfect solution for fast coral growth. These ancient colonies grew quickly over and among other corals and sponges to produce the massive, three-dimensional mounds of calcium carbonate and life that is a coral reef. When a reef like this was alive, however, its colors, swirling fish schools, swaying sea fans, whips, anemones, and the tens of thousands of other species would have made it infinitely more vivid than this carbonate gravestone. Nevertheless, the remarkable diversity of fossil types found in the lower parts of the Classic Face revealed the ancient story of a significant animal achievement.

As the crew slowly ascended the hard white walls, going up layer by layer through a long ridge covered with climbable talus, the blocky limestone changed. They had climbed through a crowded diversity of corals, brachiopods, bivalves, gastropods, bryozoans, hydrozoans, crinoids, and the many other invertebrates now turned to stone—but now the rock lost its fossil multitudes. Soon, evident in the hard limestone were only the fossils of stromotoporoids, whose hard skeletons were made up of 1- to 3-foot-wide hemispheres of white limestone, like smaller, stony bath sponges, or even like long twiggy branches of the same hard material. The rock changed color too, and soon became curiously devoid of fossils at all. It was still hard limestone, but almost featureless when the crew cracked it open with their small sledgehammers. White, granular. Empty.

At midday they finally arrived at the top of the Classic Face. The river now lay far below them, a meandering darkness bisecting this white stone valley, its crocs now too small to be visible. From there the geologists could see far to the west, but there was only the flatness of the scrub vegetation of wispy, stunted eucalyptus and the occasional squat outline of the ridiculous-looking baobab trees. Some 370 million years ago, in that direction had lain the deep ocean. To the east the crew saw another long flatness of trees and grass, but several miles farther away they could discern low hills, once the edge of the continent, with this particular reef marking the edge of the continental shelf, the last place still attached to the continent before the steep drop into the deep sea that sat in front of the Barrier Reef.

With their gear assembled, the team quickly began their measurements. The results were anticlimactic, reinforcing what the previous if sparser analyses of these rocks had quite shockingly showed: that 365 million years ago, this part of the world's largest ocean was rife with bacteria. The water would have been a deep purple because of the untold numbers of a bright purple bacterium that required sunlight and the toxic compound hydrogen sulfide (H_2S). The bacteria also had one more requirement: they could not survive oxygen in the water. It was rare in geological time to find seawater lacking oxygen at its surface; it was rarer still to have that same water saturated with H_2S in its shallow depths. But the data the team had obtained were unmistakable. Their fancy machine had chewed up a bit of ancient sediment from the Devonian reef, which, like all reefs, could exist only in water shallow enough for photosynthesis to occur. The compound they found was an unmistakable biomarker of Purple Sulfur bacteria. These forms were rare—and they were always found at times of mass extinction in the oceans over the past 500 million years of Earth history.

The geologists readied for the return trip. Before leaving, one of them took a photo that made plain the demarcation from the animal reef to the bacterial reef. The shocking part was that there was no transitional rock, no indication that this process had taken millions of years, or even hundreds of thousands of years. A catastrophe had occurred, and it had happened fast.

OCEAN EXTINCTION IN THE PAST

In the rest of this chapter I will support a contention that within several millennia (or less) the planet will see a changeover of the oceans from their current "mixed" states to something much different and dire. Oceans will become stratified by their oxygen content and temperature, with warm, oxygen-free water lining the ocean basins. Stratified oceans like this in the past (and they were present for most of Earth's history) have always been preludes to biotic catastrophe.

Because the continents were in such different positions at that time, models we use today to understand ocean current systems are still crude when it comes to analyzing the ancient oceans, such as those of the Devonian or Permian Periods. Both times witnessed major mass extinctions, and these extinctions were somehow tied to events in the sea. Yet catastrophic as it was, the event that turned the Canning Coral Reef of Devonian age into the Canning Microbial Reef featured at the start of this chapter was tame compared to that ending the 300 million– to 251 million–year-old Permian Period, and for this reason alone the Permian ocean and its fate have been far more studied than the Devonian. But there is another reason to concentrate on the Permian mass extinction: it took place on a world with a climate more similar to that of today than anytime in the Devonian. Even more important, it was a world with ice sheets at the poles, something the more tropical Devonian Period may never have witnessed.

For much of the Permian Period, the Earth, as it does today, had abundant ice caps at both poles, and there were large-scale continental glaciations up until at least 270 million years ago, and perhaps even later.[4] But from then until the end of the Permian, the planet rapidly warmed, the ice caps disappeared, and the deep ocean bottoms filled with great volumes of warm, virtually oxygen-free seawater. The trigger for disaster was a short-term but massive infusion of carbon dioxide and other greenhouse gases into the atmosphere at the end of the Permian from the spectacular lava outpourings over an appreciable portion of what would become northern Asia. The lava, now ancient but still in place, is called the "Siberian Traps," the latter term coming from the Scandinavian for lava flows.

The great volcanic event was but the start of things, and led to changes in oceanography. The ultimate kill mechanism seems to have been a lethal combination of rising temperature, diminishing oxygen, and influx into water and air of the highly poisonous compound hydrogen sulfide. The cruel irony is that this latter poison was itself produced by life, not by the volcanoes. The bottom line is that life produced the ultimate killer in this and surely other ancient mass extinctions. This finding was one that spurred me to propose the Medea Hypothesis, and a book of the same name.[5] Hydrogen sulfide poisoning might indeed be the worst biological effect of global warming. There is no reason that such an event cannot happen again, given short-term global warming. And because of the way the sun ages, it may be that such events will be ever easier to start than during the deep past.

How does the sun get involved in such nasty business as mass extinction? Unlike a campfire that burns down to embers, any star gets ever hotter when it is on the "main sequence," which is simply a term used to described the normal aging of a star—something like the progression we all go through as we age. But new work by Jeff Kiehl of the University of Colorado shows that because the sun keeps getting brighter, amounts of CO_2 that in the past would not have triggered the process result in stagnant oceans filled with H_2S-producing microbes. His novel approach was to estimate the global temperature rise to be expected from carbon dioxide levels added to the energy hitting the earth from the sun. Too often we refer to the greenhouse effect as simply a product of the gases. But it is sunlight that actually produces the heat, and that amount of energy hitting the earth keeps increasing. He then compared those to past times of mass extinctions. The surprise is that a CO_2 level of 1,000 ppm would—with our current solar radiation—make our world the second hottest in Earth history—when the five hottest were each associated with mass extinction.

MASS EXTINCTIONS

In the deep history of our planet, there have been at least five short intervals in which the majority of living species suddenly went extinct. Biologists are used to thinking about how environmental pressures slowly choose the organisms most fit for survival through natural selection, shaping life on Earth like an artist sculpting clay. However, mass extinctions are drastic examples of natural selection at its most ruthless, killing vast numbers of species at one time in a way hardly typical of evolution.

In the 1980s, Nobel Prize–winning physicist Luis Alvarez, and his son Walter Alvarez, first hypothesized that the impact of comets or asteroids caused the mass extinctions of the past.[6] Most scientists slowly come to accept this theory of extinction, further supported by the discovery of a great scar in the earth—an impact crater—off the coast of Mexico that dates to around the time the dinosaurs went extinct. An asteroid probably did kill off the dinosaurs, but the causes of the remaining four mass extinctions are still obscured beneath the accumulated effects of hundreds of millions of years, and no one has found any credible evidence of impact craters. Rather than comets and asteroids, it now appears that short-term global warming was the culprit for the four other mass extinctions. I detailed the workings of these extinctions first in a 1996 *Discover* magazine article,[7] then in an October 2006 *Scientific American* article, and finally in my 2007 book, *Under a Green Sky*.[8] In each I considered whether such events could happen again. In my mind, such extinctions constitute the worst that could happen to life and the earth as a result of short-term global warming. But before we get to that, let us look at the workings of these past events.

The evidence at hand links the mass extinctions with a changeover in the ocean from oxygenated to anoxic bottom waters. The source of this was a change in where bottom waters are formed. It appears that in such events, the source of our earth's deep water shifted from the high latitudes to lower latitudes, and the kind of water making it to the ocean bottoms was different as well: it changed from cold, oxygenated water to warm water containing less oxygen. The result was the extinction of deep-water organisms. Thus a greenhouse extinction is a product of a changeover of the conveyor-belt current systems found on Earth any time there is a marked difference in temperatures between the tropics and the polar regions.

ANATOMY OF GREENHOUSE EXTINCTION

Let us summarize the steps that make greenhouse extinction happen. First, the world warms over short intervals due to a sudden increase in carbon dioxide and methane, caused initially by the formation of vast volcanic provinces called flood basalts. The warmer world affects the ocean circulation systems and disrupts the position of the conveyor currents. Bottom waters begin to have warm, low-oxygen water dumped into them. The warming continues, and the decrease of equator-to-pole temperature differences brings ocean winds and surface currents to a near standstill. The mixing of oxygenated surface waters with the deeper and volumetrically increasing low-oxygen bottom waters lessens, causing ever-shallower water to change from oxygenated to anoxic. Finally, the bottom water exists in depths where light can penetrate, and the combination of low oxygen and light allows green sulfur bacteria to expand in numbers, filling the low-oxygen shallows. The bacteria produce toxic amounts of H_2S, with the flux of this gas into the atmosphere occurring at as much as 2,000 times today's rates. The gas rises into the high atmosphere, where it breaks down the ozone layer. The subsequent increase in ultraviolet radiation from the sun kills much of the photosynthetic green plant phytoplankton. On its way up into the sky, the hydrogen sulfide also kills some plant and animal life, and the combination of high heat and hydrogen sulfide creates a mass extinction on land.[9]

Could this happen again? No, says one of the experts who write the RealClimate.org Web site, Gavin Schmidt, who, it turns out, works under Jim Hansen at the NASA Goddard Space Flight Center near Washington, DC. I disagreed and challenged him to an online debate. He refused, saying that the environmental situation is going to be bad enough without resorting to creating a scenario for mass extinction. But special pleading has no place in science. Could it be that global warming could lead to the extinction of humanity? That prospect cannot be discounted. To pursue this question, let us look at what might be the most crucial of all systems maintaining habitability on Planet Earth: the thermohaline current systems, sometimes called the conveyor currents.[10]

CONVEYOR CURRENTS IN THE PAST

It is both presumed and observed that current systems that run like a conveyor belt (it runs horizontally until ducking down, reversing direction, and returning up to its original starting point) are among the most important of the many ways that the earth redistributes heat from the sun. Such current systems have been present on Earth whenever there has been ice at the poles, and perhaps when there is no ice at all. In the past, short-term global warming caused perturbations to several of the conveyor current systems. Will the melting of Greenland and Antarctica cause such perturbations in the near, warmed future? Could these changes even be happening now? And if so, what might the consequences be?

Today the most important of these currents appears to be the one that moves warm water north and east from the warm Gulf Stream of eastern North America. As that current moves into higher latitudes, its water cools and finally sinks. This cold, highly oxygenated water is a crucial part of maintaining a mix among the ocean's gaseous elements, rather than allowing them to become stratified, with oxygenated tops and oxygen-free bottoms, like today's Black Sea, or even totally anoxic from bottom to top. If the Gulf Stream–related current were to change the position where the water sinks, so that less-oxygenated warm water sinks from the surface or so that no water sinks at all, which would be the cessation of the current system, Europe might be immediately cooled, even in a globally warmed world, at least for a while. The result would certainly be a great change in the weather, which would certainly affect agriculture, and probably not for the better. In 2005, for the first time, a research group reported a slowing of the North Atlantic conveyor current, probably due to massive amounts of freshwater already entering the sea in northern areas due to the rapid melting of the northern ice cap.[11] As this melt increases in volume, the current will be massively affected. Freshwater is of lower density than seawater, and it will float along the top of the ocean, effectively stopping the conveyor action of the current itself.

Just how sensitive is the conveyor current to the sort of change that could lead to a major disturbance in the world's climate—the kind of dramatic global change that in the past caused mass extinction? In other words, what would it take to cause a short-term but radical change in the conveyor current? Some climatologists regard the Atlantic current as robust; they believe that only massive changes in oceanography would be required to perturb it. But a larger number of scientists, including Richard B. Alley, in his now classic and important 2002 book *The Two-Mile Time Machine*, regard the Atlantic conveyor current system as very finely balanced and hence very susceptible to change.[12] The easiest way to activate this change, according to sophisticated computer models, is to pump freshwater into the northern part of the system, and that is just what is happening today. The truly staggering rate at which Arctic ice is melting—a phenomenon not even noticed before about 2003—is introducing massive volumes of freshwater into the most dangerous point for the integrity of the conveyor current. And that input of freshwater is really just the tip of the melting iceberg. However, another way to change the system is by rapid global temperature rise, of sufficient magnitude to significantly reduce the temperature difference between poles and equator.

The consequence of perturbation to this system is that the deep, cold, and oxygenated bottom water from high-latitude sinking will change to deep, warm, anoxic water that came from mid-latitude sinking. With that change a relatively cool world gives way to worldwide tropics. But could this happen again and if so, how soon? These questions stimulated an interesting NASA meeting in 2009.

THE 2009 AMES MEETING

In January 2009 I received an unexpected telephone message from Dr. Carl Pilcher, director of the NASA Astrobiology Institute (NAI), summoning me to a spring meeting of the NASA Ames Research Center in Sunnyvale, California, to join a discussion on life and planetary change. It turned out that the director of Ames, former astronaut Pete Worden, had instigated the meeting to discuss the implications of short-term climate change on global biodiversity, past and present. Greenhouse extinction, in other words.

Thus a small group composed of scientists who have each worked on either past mass extinctions or on the consequences of ancient climate change convened in welcome California warmth. We were all glad to meet with NASA, because it had been frustrating to see how little traction this concept had gotten with the public, other scientists, and the national agencies that fund scientific research. The other scientists attending were fellow paleontologist Doug Erwin of the Smithsonian; geochemists Lee Kump of Penn State and Ariel Anbar of Arizona State; biologist Jon Harrison, also of Arizona State; biochemist Roger Summons of MIT; and climate modeler Jeffrey Kiehl of Colorado.

In making our presentation to a small cadre of NASA scientists and administrators, Summons, Erwin, and I conveyed data and information supporting the hypothesis that more than one of the past mass extinctions might have been caused by short-term global warming, with the devastating Permian mass extinction especially featured.

Next, several scientists reported about the prospect of future greenhouse extinctions. Lee Kump of Penn State spoke first, and that was highly appropriate, for in 2005 he and colleagues first published the evidence suggesting that H_2S played a major role in mass extinction. Kump showed the results of modeling of the Atlantic and Pacific oceans that investigated whether the gigantic thermohaline conveyor currents (integral to keeping the deep ocean oxygenated) could soon be affected by polar warming and the infusion of freshwater. He also added a new and important variable: the effect of enhanced nutrients to the deep ocean at the same time as global warming. He included this factor because the mechanism that he proposed for the Permian extinction, while triggered by global warming, had as its real "kill mechanism" the formation of vast quantities of hydrogen sulfide dissolved in the oceans (and at high enough concentrations, leaking into the atmosphere, literally bubbling out of the sea). For H_2S to be produced by microbes from a group that used sulfur, not oxygen, for respiration, and to get large enough quantities of H_2S to kill things, there would have to be a lot of nutrients down there. To my surprise, his findings indicated that both the Atlantic and Pacific oceans could see the start of oceanic slowdown not millennia hence, but early in the next century. The only factor in his scenario that Kump failed to take into account was rising sea level. It is this mechanism, perhaps more than any other, that would put the necessary nutrients onto the bottom of the sea, for as the many rivers and river mouths drowned, vast quantities of organic-rich silt and mud would be carried out to sea, where it would serve as fertilizer, rich in phosphates and nitrates that could stimulate the growth of the H_2S-producing microbes, akin to fertilizing a garden bed filled with plants producing deadly poison.

Jeffrey Kiehl, who was the day's final presenter, also used models to look into the near future. He too saw signs that changing oceans are heading toward low oxygen and that warmed ocean bottoms could begin in the current century if current global warming persists. Yet in all of the models he neglected the topic of this book: the effects that rising sea level will have on global temperatures. Water absorbs heat from the sun and generally reflects back into space less energy than land surfaces do. Thus, all else being equal, the larger the ocean area, the greater the warming through reduced albedo (planetary reflectivity). It is a vicious circle, a positive feedback. Snow and ice melt, reducing albedo and raising sea level. As the sea rises, it absorbs ever more heat, causing more ice to melt at the poles, again raising sea level, and on and on.

The result of this Ames meeting was a report that NASA said was headed to the desk of President Barack Obama's science adviser. Whether it got there we never found out. But what we do know is that NASA has seemingly awakened to the vital connection between ancient climates and impending climate change.

Although a number of scientists have tried to communicate this argument to the public, at the end of the first decade of the new millennium, few in the nation's electronic media and print and newspapers allowed us to make our case. They did not disbelieve us; they just responded that the past scenarios were too horrifying for us to contemplate that they could happen again, and soon. Let us hope that a new generation will quickly decide to open their ears and listen.

CHAPTER 8

STOPPING CATASTROPHIC SEA LEVEL RISE

The earth from 200 miles up, 3200 CE.
Carbon dioxide at 550 ppm and dropping.

From space, the earth has taken on a new appearance over the past 1,000 years. The giant orbiting mirrors kept wide swaths of the planet's surface in shadow, giving the sunlit half of the globe a mottled appearance. All of the coastal areas of the oceans, and huge regions in the middle of the oceans, were now green. The immense quantities of iron filings, so laboriously dumped into the seawater, had done their job: great pastures of phytoplankton now filled ocean environments that had long been the home to little life. As those single-celled plants proliferated, they sucked up carbon dioxide from the sky. Viewed from space, the earth offered other, smaller bits of evidence indicating that major technological changes were under way. All of the countries of the mid-latitudes showed immense fields of black: cheap solar cells had finally been developed, and when the sun's rays were allowed to reach various areas of the earth's surface, the enormous panels created electricity. Giant pipes could be seen extending from many parts of the coastal oceans, with fine mist spewing into the atmosphere. Around every city a cordon of windmill-sized carbon dioxide scrubbers removed the industrial CO_2 wastes that urban areas so readily produced.

There was something even more different about the night side of the planet. It no longer looked like a light-infested Christmas tree on steroids. The swaths of light that had formerly extended down the entire eastern and western seaboards of North America, as well as through most of Asia and Europe, and driven astronomers so mad during the late twentieth and early twenty-first centuries, were now greatly dimmed. The giant cities still bejeweled the earth's dark side, but they were smaller, single diamonds of light, rather than the garish snarls of costume jewelry that had arisen as cities begat suburbs, and suburbs metastasized along the crisscrossing freeways. The cities had pulled themselves in; many of the overbuilt spiderwebs of roadage had fallen away. Nighttime was now an occasion to embrace darkness.

The planet's coastlines bristled with defense against the rising seas—dikes, but mainly offshore sandbars and gates in front of estuaries. But all of these barriers had now become superfluous. The sea had stopped rising. The ice had ceased to melt. The final tally of the encroachment was a 6-foot increase. That was enough to cause damage and death, but it fell far short of the catastrophe it could have been. The truce with the oceans had come at a cost, however. People no longer traveled so freely. Governments levied enormous fines for single drivers in cars. Distance driving was heavily taxed.

The restrictions and the new technology had worked—the combination of voluntary and government-enforced emission reduction, along with geoengineering—planetary-scale projects designed to deal with the many aspects of climate change and changing planetary habitability—on a vast scale, had combined to lessen the upsurge in carbon dioxide and the melting of polar ice. Neither effort could have effected the change alone. The enterprise had seemed foolhardy at first, the restrictions unnecessarily onerous; some considered the hundreds of billions of dollars spent on geoengineering solutions to be a gigantic boondoggle. But gradually the coastal cities were saved. The tipping point that would have capsized the planet into utter catastrophe was not reached.

At the planet's extreme north and south, the two most beautiful of all of Earth's jewels shone brightly. Greenland and the Arctic sea to the north and Antarctica and its ice sheets in the south reflected sunlight back into space—light that began a reflected voyage into the farthest reaches of the galaxy, where other life might exist.

Far from Earth, a hundred light years away, in fact, another intelligent race looked at the imaging they had made of the distant planet. They saw an orb with artificial lights, but one whose ice caps neither grew nor shrank, and they knew they had found another truly intelligent form of life.

IS THERE HOPE?

The scenario above, the penultimate of this book's travels into the future, offers us the best we could hope for. I trust that it also conveys the enormity of the changes in human behavior and technology required for us to avoid disastrous climate change. Without such profound technological advances and alterations in behavior, we veer toward the other extreme, with the earth growing so warmed that not only do we lose the ice caps but we also ensure a mass extinction. That does not have to happen. Even though we have a long way to go, progress has been made—at least in our level of awareness of the problem. A decade ago, climate change was not news. Now it is part of our consciousness. It is the rare human on this planet who does not have a view one way or another about the changes in climate, and that is the best news. Unfortunately, the news about the climate itself is usually bad; we learn constantly of rising temperatures, rising carbon dioxide levels, and rising seas. Yet amid all the bad reports, short flashes of hope appear from time to time, in ways and places often subtle—but enough to indicate the opportunity of a world where global warming is arrested; where the seas do not rise from their basins; and where humans actively reduce their numbers as well as their greenhouse gases. It would be wonderful to live in a world in which the only thing going up was the global standard of living. We can get there only if we still have hope. Hope is not only a motive, but from where we stand now, perhaps itself a goal. Maybe we are at the stage where our best efforts will lead us to be hopeful, because we are actively doing things.

Is hope realistic? I will finish this not overly cheerful book with some specific strategies that, if successfully employed, could indeed give us hope that the ice sheets will not uncontrollably melt and that the seas will not catastrophically rise. But unless we change our attitudes and engineer new climate-protecting technologies, the very possibility of such strategies is itself problematic. While I am sure that, barring major change, we are heading toward what indeed will be a flooded world, I am not sure that we have the will to do what is necessary. Over the course of writing this book, I found myself not a little shaken by all I had learned and synthesized. I grew increasingly pessimistic about the prospect of forestalling calamity. At the same time, I realized that offering a downer of an ending could indeed be counterproductive to my cause—which is to encourage the reader to try to change the world.

While I was completing this book, I met with the Australian environmentalist Tim Flannery, who wrote the best-selling *The Future Eaters: An Ecological History of the Australasian Lands and People,*[1] as well as many other eloquent and important works about climate change. Flannery is now part of the group trying to "save the world" through the Copenhagen treaty—an effort by environmentalists worldwide to create a blueprint for international action on emissions reductions.[2]

Called the "Copenhagen meeting" by all concerned, the meeting was really named "COP15," itself an acronym for the 15th Conference of Parties, or countries, to the UN Framework Convention on Climate Change (UNFCCC). COP15 is also the fifth meeting of parties to the Kyoto Protocol, a legally binding emissions-reduction treaty created in 1997 in Kyoto, Japan. The Kyoto agreement aims to reduce global industrial greenhouse gas emissions by an average of 5 percent against 1990 levels over a five-year period, from 2008 to 2012. The Kyoto climate treaty, which went into force in 2005, was ratified by 185 nations—but not the United States. Because the Kyoto Protocol expires in 2012, an "ambitious new deal" needs to be worked out in 2010 to provide governments guidance beyond Kyoto, the UNFCCC says, and hence Copenhagen.

Tim Flannery was one of the thousands going there, with the best of intentions, to literally try to save the world—at least as we know it. Yet even when we met, prior to the late-2009 conference in the eponymous city, I could see that he had high hopes but far more realistic expectations for some sort of global agreement. Unfortunately, at its conclusion the ultimate result of Copenhagen seemed to have been a failure of even its most modest goals.

Going in, the goals of that conference seemed clear enough. They consisted of:

1. Make clear how much developed countries, such as the United States, Australia, and Japan, will limit their greenhouse gas emissions.
2. Determine how, and to what degree, developing countries, such as China, India, and Brazil, can limit their emissions without limiting economic growth.
3. Explore options for "stable and predictable financing" from developed countries that can help the developing world reduce greenhouse gas emissions and adapt to climate change.
4. Identify ways to ensure developing countries are treated as equal partners in decision-making, particularly when it comes to technology and finance.

The ultimate goal was and is to reduce emissions.

What of the results? Unfortunately, after more than a week of sniping, press leaks, and huffy entrances and exits, the results were minimal. The Copenhagen Accord set no goal for concluding a binding international treaty, leaving months, and perhaps years, of additional negotiations before it emerges in any internationally enforceable form. The only tangible result was that the conference appeared to have caused money in notable quantities to start flowing . . . from rich nations to poorer ones.

A REALITY CHECK

We too can focus on tasks that are immediate and doable, even as we examine some ambitious and far-reaching solutions. We are faced with three possibilities. The first posits that all or most of the interpretations and conclusions of so much recent climate science—that rising carbon dioxide will lead to catastrophic sea level change in the not-so-distant future—are wrong. Under this supposition, carbon dioxide really has *no* effect on global climate, or it does but the ice sheets will not melt no matter how much greenhouse gases rise. Or, in a variant on this possibility, perhaps the world really does warm significantly but the ice either does not melt at all or does so in such limited amounts, or so slowly, that humans do not experience a significant environmental problem. We might discover that it takes a *lot* longer to melt ice sheets even at higher-than-normal temperatures than this book suggests. This prospect contends that we could do nothing to counter whatever climate changes occur—and get away with it. In this event, climatologists would be perceived as Chicken Littles, their dire warnings the object of mirth from every climate-change denier who will cry, "I told you so."

The second possibility is that the ice melts significantly, seriously, quickly, and dangerously—and we do something about it, with such alacrity and efficacy that we stall the sea in its tracks and avert chaos in the world economy, uproar in global society, and massive human mortality. With this outcome, it turns out that we are indeed able to "fix" things through some combination of emissions reduction and geoengineering to stabilize sea level in a manner favorable to civilization. And maybe the planet will cooperate with our efforts, with climate change coming more slowly and modestly than predicted in this book, thus allowing us to manage its effects. If we are very optimistic about human ingenuity and adaptability, maybe even a 6-foot rise in sea level is something we can live with. (Personally, I would rejoice at the certainty of a mere 6-foot rise.)

Then there is the third possibility, the one that as a scientist I believe is the most likely: the ice sheets melt rapidly, the sea rises ferociously, and all the scenarios set forth in this book come to pass.

HUMAN ACTIONS TO FORESTALL CLIMATE CHANGE

Confronted with these three quite different prospects in mind, what do we do? As this book's arguments and evidence make extremely clear, I reject the possibility that climate science is so far off the mark there will be no rapid global temperature rise—and thus no flooded world. Which leaves us with possibilities two and three, and number three just cannot be allowed to happen. In any event, if we significantly reduce pollution it will be only because we have produced a cleaner world, with more efficient industry, and diversified energy sources replacing most coal and oil—and because we have raised living standards everywhere.

Yet even if humanity gets its act together and, through behavioral as well as technological change, reduces emissions to the point that the ice sheets do not melt, I believe our efforts will turn out not to be enough. It is too late to forestall climate change entirely through merely reducing greenhouse gases. We will have to significantly change our sources of energy and the means of delivering it. We will also have to undertake vast engineering projects (both mechanical and biological, such as wholesale tree planting and other botanical sinks for CO_2) that, when combined with more individual efforts, will be enough to tip the scale in our favor. Those efforts could save the ice sheets, even as the sea rises the 3 feet already anticipated or even 6 feet (sorry, Holland and Bangladesh) and then stops. To accomplish this feat, we need to limit global warming to perhaps no more than 3 or 4 degrees Fahrenheit.

Why is that goal important, and how realistic is it? A 2006 column at RealClimate.org, which I consider the best public source for information on climate change, offers a compelling summary of this problem as well as a possible solution.[3] Authors Malte Meinshausen, Reto Knutti, and Dave Frame went through the math showing that a stable CO_2 level of 400 ppm yields an 80 percent chance that the earth will warm no more than 3 to 4 degrees Fahrenheit. For instance, the rise from CO_2 levels of 280 ppm at the start of the Industrial Revolution to the present 390 ppm level has brought about a 1- to 2-degree Fahrenheit global temperature rise, thus providing an independent test of the climate models used to predict future temperature rises that are tied to rises in greenhouse gas concentration. The good news is that methane, one of the most troublesome of greenhouse gases now being produced by human activity, has a short life in the atmosphere before it breaks down. Moreover, the oceans are an effective sink for atmospheric carbon, as we have seen in a previous chapter. If human emissions can be sharply curtailed soon, then concentrations of all greenhouse gases could begin to decline near the end of the century, although as noted above it would take tens of thousands of years to remove all of these pesky carbon dioxide molecules out of the atmosphere. The RealClimate.org model even allows greenhouse gas levels to peak at 475 ppm for a short time, but we do not go past the 4-degree Fahrenheit rise if we can then bring them back down to 400 ppm before the end of this century.

We could make that hopeful model a reality by using beneficial new technology (and in some cases by abandoning some toxic old technology) and through hard work and common sense.

We can start with the easy stuff, the things we each have control over. We can drive less. We can drive less-polluting cars and buy hybrids or electric cars. We can educate ourselves on what it is we are eating, wearing as clothing, living in, buying, investing in, and voting for, and the effect of each of these factors on our climate. Then comes the real challenge: making structural and way-of-life changes on a global scale. We will need to stop using coal to the extent that we can. We will have to employ solar, tidal, wind, and even nuclear power. Beyond altering how we make and use energy, we will have to reengineer the biological world. Vast regions deforested over the past century will need to be reforested. One way to help the land to literally recover itself is to plant fast-growing weed biomass, then char it, and mix the elemental carbon in large quantities deep into soil in tropical regions where a combination of already-thin soil with heavy, soil-eroding rainstorms has stripped the land down to bedrock. This process conditions the soil and as a bonus reduces conventional fertilizer need by 50 percent.[4] But perhaps even more important, we must transform farming practices radically. One of the greatest threats to the world food supply is the loss of topsoil through deleterious agriculture. Even slight sea level increases will reduce world crop yield; we need to counteract that decline by saving topsoil and also reducing the practice in tropical regions of stripping away forests for short-term livestock husbandry on soil that soon erodes away.

A challenge perhaps even more difficult to meet will be our voluntary abstention from our most cherished technological and social accomplishments, and the freedom that comes with them. I am not referring to political freedoms, although those too will have to be infringed upon to enact the necessary changes. The most stringent of these will have to do with transportation. The vaunted "freedom of the open road" will have to disappear. The use of private vehicles for all but the most vital tasks will have to be abandoned. So too will air travel be revamped and restricted. Unless we replace our fast jets with new forms of large dirigibles, there is no way we can afford to fly at will—jet planes are prodigious greenhouse gas emitters. Sailing ships, well rigged with fast communication facilities, might replace the polluting steamships for both travel and trade. We will have to revive the archaic luxury of an ocean voyage, and that will really not be so bad. In a globally connected world, taking a week to get to a business meeting overseas instead of an overnight flight might actually be a lot more fun than the current madness of airports and airport security.

Consonant with curtailing transportation will be changes in how we occupy our living space. Cities will have to become more condensed: we must lose the suburbs and the shopping malls outside of cities. We must be able to work at home or near home and meet our daily needs without using personal automobiles. We must reduce military activities; the military materiel, training, and deployment of armies, navies, and air forces produce unknown quantities of greenhouse gases. And most important, we must reduce human population. There are too many of us on the planet as it is, and as we've seen, there threaten to be billions more. I recognize that reducing our numbers is a fraught endeavor, socially and politically. Paradoxically, it will happen only when living standards are high and equitable for all peoples. Safely raising those standards while at the same time reducing emissions will be the greatest challenge of all.

TECHNOLOGICAL FIXES

I do not think that even the draconian list of social and behavioral changes outlined above will be enough to keep us from catastrophic sea level rise. Forestalling the rising of the sea, I suspect, will require engineering on a global scale of unprecedented complexity and expense. I am not alone in believing this. A huge debate rages about the possibility, the necessity, and even the wisdom of attempting large-scale technological fixes to ameliorate climate change. These efforts are generally referred to as geoengineering.

The concept of geoengineering received a hugely visible lift with the publication of *SuperFreakonomics: Global Cooling, Patriotic Prostitutes, and Why Suicide Bombers Should Buy Life Insurance*[5] by Steven Lefitt and Stephen Dubner, who also wrote the earlier best seller *Freakonomics: A Rogue Economist Explores the Hidden Side of Everything*.[6] In a whole chapter of their second book, these two intrepid economists strongly advocate for two different methods of geoengineering.[7] The first is to shoot sulfur into the atmosphere, the proposal earlier (and controversially) made by Nobel Laureate Paul Crutzen. The idea has merit, the authors say; sulfur has a cooling effect, as can be observed after a large volcanic eruption. The second method Lefitt and Dubner advocate is creating artificial cloud cover, by modifying clouds to be more reflective, which would change the earth's albedo (global reflectivity) and cause more sunlight to bounce back into space. But what about oceanic acidification? That problem is not going to go away anytime soon, and it highlights a biological problem that is just not conducive to the optimistic fixes advocated by the two best-selling authors. Already small mollusks called pteropods are having the calcareous shells eaten off their backs because of high oceanic acidity in the Arctic seas. These tiny animals are the base of a food chain that itself could collapse with the absence of the pteropods, because loss of shell kills the animals.

Although *Superfreakonomics* is flashy, it is overshadowed by the more sober and scientific examination of the most useful entrée into the discussion, a 2009 report by a blue-ribbon panel of British scientists, the Royal Society of Great Britain.[8] The report presents geoengineering as a perhaps necessary but less-than-ideal option for reducing global warming. As realists, the authors of the report stress that nothing should divert us from the main priority of reducing global greenhouse gas emissions in our own personal lives and that to do this requires effective action to reduce emissions of greenhouse gases. They are saying that it might be all too easy to put too much confidence in global rather than local solutions. The authors offer no hope for a technical panacea. Geoengineering methods cannot replace climate change mitigation, and we must view them as part of a larger global effort. The greatest danger is "falling in love" with technological fixes as a substitute for real changes in reducing emissions.

The report proposes that geoengineering efforts be broken up into those relating to managing energy coming from the sun (they termed this Solar Radiation Management techniques) and those dealing with the amount of carbon dioxide in the atmosphere (Carbon Dioxide Removal techniques). Within these two kinds of technical projects, a variety of plausible geoengineering "solutions" were explored and assessed: biochar, enhanced weathering, carbon dioxide air capture, ocean fertilization, surface albedo alterations (in both urban and desert environments), cloud albedo modification, stratospheric aerosols, and space reflectors. They were evaluated on their effectiveness, affordability, timeliness, and safety. Interestingly, although some of the report's projects might illustrate the cover of a 1950s science fiction pulp magazine, such as giant space reflectors, others are more homely, such as putting charcoal into soil, which can be done by hand, on a small scale, by anyone.

The Royal Society report helps sort out these various options with a useful graph comparing the effectiveness of each against its relative cost. In this section the figures quoted are all from the report. As might be expected, no geoengineering project comes cheaply. Some of the solutions could even bankrupt any single country. And some possible solutions that have received praise in other circles are given short shrift in the report, schemes such as painting all houses and building roofs white to reflect sunlight back into space, or increasing carbon sequestration by burying charcoal. Interestingly, the report favors removing carbon dioxide over limiting the amount of sunshine hitting the earth.

Perhaps the most powerful of all proposed geoengineering methods involves biology much more than either geology or engineering. The day-to-day biological effects of terrestrial ecosystems are very good at getting rid of carbon. For instance, perhaps as much as 3 billion tons of carbon are removed each year from the atmosphere through the living processes of forests, fields, and other non-oceanic ecosystems. For us humans, the simple act of living by a forest removes our carbon emissions from the atmosphere, and the overall effect is to absorb about a third of CO_2 emissions from fossil fuel burning and net deforestation. In fact, the world's forest ecosystems store more than twice the carbon that exists in the atmosphere. Hence, keeping forest and other plant-rich communities healthy does much to reduce atmospheric CO_2. But every time we cut down a forest, all of that natural carbon-removal machinery disappears. This is a huge problem: currently; as much as 20 percent of emissions caused by humans comes from changes in land use, mainly deforestation. The December 2009 Copenhagen meeting on climate change may have failed to do much to halt the threat of rapid sea level rise, but many nations at least agreed that deforestation is a serious problem and that provisions to pay poorer countries to preserve their forests are critical. Even a single tree can make a tree cutter a lot of money, and yet every tree is a powerful weapon against too much CO_2.

Tropical deforestation alone now accounts for about 16 percent of global emissions and is the fastest-rising source of emissions. Locally, better land-use management that incorporates carbon storage (such as in underground depositories) would help greatly. Globally, nothing will help more than reforestation following reduced deforestation—a net increase in trees over the present day. The problem, of course, is money. Who will pay to stop cutting and then replant already-cut forests? At Copenhagen this was a large part of the annual $100 billion U.S. secretary of state Hillary Clinton promised for the world's poorer tropical countries to reduce deforestation and increase reforestation. If we use biology itself to profoundly change the makeup of the atmosphere simply by planting lots of trees, who is going to police the forests? I remember being on the Pacific island nation of Vanuatu in 1986 and seeing gigantic Japanese ships that had been engineered to move right onto the shore to remove gigantic hardwood trees from the island. It reminded me of the many films showing Japanese whaling practices, where the still-twitching whales are hauled into the maw of a giant factory ship. Both whaling and wholesale logging—for the valuable but slow-growing hardwoods—occur out of sight of society. Effective implementation of anti-deforestation and pro-reforestation measures will depend on reliable baseline estimates, monitoring, and enforcement.

Another land-based method for reducing carbon levels is burying charcoal. By putting organic carbon back in the soil in the form of charcoal, a great deal of carbon is removed from the carbon cycle, reducing the amount in the atmosphere. For example, it has been proposed to bury wood and agricultural waste both on land and in the deep ocean to store carbon rather than allow decomposition to return it to the atmosphere. But any such methods require energy—energy to bury the enormous volumes of charred wood needed to make a difference, and energy to move all of this material to appropriate geographic sites. Moreover, the processes may disrupt the involved ecosystems in profound ways—we may wind up managing certain ecosystems to death, causing more harm than good. In the deep ocean, for example, organic material would be decomposed and the carbon and nutrients returned to shallow waters. This could perturb the ecosystem in ways we are far from understanding. Full assessments are not yet available for the costs and benefits involved.

Another land-based carbon dioxide management method would be to grow plants that reflect more light back into space—plants that are light green rather than dark green in color. Yet another technique that is not biologically driven would be to build large reflective surfaces on deserts. While the technology involved is feasible, the result would be major ecological uproar, not to mention the carbon cost of whatever reflective material was used. Nevertheless, there is a lot of desert area on Earth today, with more each day due to desertification. The ecological balance sheet of such technology would be problematic at best. A lot of lizards, turtles, insects, birds, and cactus would die, but animals in other areas would gain a reprieve. Such choices—always choices—will be difficult.

A final tree-based solution, suggested by former Princeton physicist Freeman Dyson, relies on future advances in biotechnology. He has predicted that within fifty years scientists will be able to genetically enhance trees' ability to "eat" carbon dioxide, and convert it into a stable form and bury it into the ground. If we replanted 25 percent of the world's forests with carbon-eating versions of each species, we would reduce the CO_2 in the atmosphere by 50 percent in fifty years. But of course we have no idea at this time about the actual possibility or ramifications of such alternations to the basic schemes of nature.

It is not only reforestation on land that makes sense but also carbon dioxide reclamation via the sea. Most of the CO_2 being released today sooner or later will make its way into the deep ocean. This process occurs on millennial time scales, and if we could speed it up the result could be significant decreases in CO_2 levels in the atmosphere. To accomplish this, we could either cool the oceans (cold water takes up carbon dioxide faster than warm water) or increase the amount of plant life in them. Although cooling entire oceans is impossible, we might be able to increase aquatic plant growth, in a way analogous to reforestation. We could "fertilize" the oceans by infusing them with nutrients, mainly iron, that would encourage the growth of plant life and increase CO_2 uptake by purely biological methods of photosynthesis. Unfortunately, recent tests of this technology have not been promising.

Theoretically, sprinkling iron compounds onto the surface of the sea should encourage the growth of phytoplankton, which takes up CO_2. When the phytoplankton die, they sink to the ocean bottom, dragging their carbon with them and out of the carbon cycle. The result is less carbon available for further plant growth, but more important, less CO_2 as a greenhouse gas. However, the results of several small-scale pilot studies of this procedure have been underwhelming. Increased iron did lead to the predicted bloom of phytoplankton, but the effect was mitigated by another biological mechanism: the novel appearance of phytoplankton in oceans previously too nutrient-starved to allow much plant growth stimulated a population explosion of phytoplankton-eating organisms, mainly small animals, which in turn were eaten by higher organisms on the food chain. In the most ambitious of these experiments, the Indian-German Lohafex expedition in early 2009 fertilized 116 square miles of the Southern Atlantic with six tons of dissolved iron. Soon after the expected proliferation of phytoplankton, the small crustaceans known as copepods moved in and scarfed up all the new phytoplankton. The copepods were in turn eaten by predators, and all of the creatures' carbon went right back into the atmosphere through normal animal respiration.

Not only biology could fight rising atmospheric CO_2. Carbon dioxide could also be removed from the atmosphere by accelerating the natural weathering process of ores. It turns out that some common rock types, most importantly rocks such as granite and basalt that are rich in silicate minerals, can actually remove CO_2 from the atmosphere during the processes of chemical weathering. This reaction consumes one CO_2 molecule for each silicate molecule that weathers, storing the carbon as a solid mineral. It is called the Silicate Carbonate Weathering cycle. It has also been called Earth's thermostat, a feedback system that has been crucial in keeping the planet's temperature at levels allowing the presence of liquid water over geological time—and thus allowing habitability. If we mined the rock, ground it to powder, and then spread the silicate-rich dust over fields, we could speed up this natural process. It is estimated that if we spread out the same volume of rock as all the coal we mined each year, we would remove as much CO_2 as we currently emit. But such an endeavor would require an immense expenditure of energy. Since the dust would be light in color (if it were mined from granite) it might even help provide us additional reflectivity—thus cooling the earth.

Alternatively, it has been suggested that carbonate rock could be processed and ground and then released into seawater, thus increasing the oceans' alkalinity, resulting in additional uptake of CO_2 from the atmosphere. This procedure would also resolve the major problem of ocean acidification.

A totally different solution to our atmosphere's oversupply of carbon dioxide is to remove it directly from the air itself. At present, two main technological routes are being pursued to develop large-scale commercial capture of CO_2 from the air: adsorption onto solids, and absorption into highly alkaline solutions. In both cases, solids of various kinds react with air to cause a chemical reaction using carbon dioxide, changing it into a different, less harmful chemical compound.

Another, perhaps obvious solution to climate change is not Earth based, but takes place in space: cutting off sunlight to the planet before it gets here. Simply put, big mirrors in space would reflect some of the sun's radiation, thus cooling the earth. By reflecting just 1 percent of sunlight back into space, the earth would cool dramatically—enough to offset the temperature increases from all the greenhouse gases generated since the Industrial Revolution. But making enough mirrors to reflect even 1 percent of sunlight would be a massive (and expensive) undertaking—expensive both in money and in yet more emissions produced to build the mirrors and send them into space. Another problem is that some of these mirrors (in the lowest orbits) could fall out of orbit and potentially crash into the earth, although most would surely burn up before hitting the surface.

There has been no shortage of ideas about what substance the mirrors would be made of. One notion is to build a giant mirror on the moon using the many tons of lunar glass existing there. But frustratingly, none of these studies take into account the carbon liberated by the vast quantities of energy required to build all this stuff. The construction of a moon rocket, let alone the manning of a lunar base, would require the burning of so much coal on Earth that all positive effects would be lost. Other possibilities include building and placing into space a cloud of trillions of thin metallic reflecting disks, fabricated on Earth and launched into space in stacks of a million, one stack every minute for about thirty years. This sounds mighty far-fetched to me.

A better possibility is a reflective space mesh. First proposed by Edward Teller (creator of the hydrogen bomb), the idea envisions a superfine reflective mesh to be engineered out in space and positioned between the sun and the earth.[9] Working on the same principle as space mirrors, the mesh would be much cheaper to construct. Furthermore, just a few years ago, Ken Caldeira from the Carnegie Institution Department of Global Ecology at Stanford University in California simulated the effects of such diminished solar radiation and found it to work well.

Yet whether we are talking mesh or mirrors, all of these space-based approaches are just plain scary. There is plenty of space junk that could cause a mirror to go off course or change the direction of its beam. And if any of the big mirrors came back down to Earth unexpectedly, disaster could result. Nevertheless, space-related solutions for climate change are technologically possible, and more and more specialists think they should be undertaken.

Finally, let us look in detail now at the sulphur scheme the *Freakonomics* guys advocate. Although controversial, this variation on the solar shield theme would be cheaper and easier to implement than alternatives based in outer space. The basic idea comes from what happens during volcanic eruptions, in which sulfate particles shooting into the stratosphere reflect radiation away from the earth, lessening the heat from the sun.[10] In theory, injecting a massive load of sulfur into the upper atmosphere would have the same effect. A wide range of types of particles could be released into the stratosphere with the objective of scattering sunlight back into space. Either hydrogen sulfide (H_2S) or sulfur dioxide (SO_2) could be introduced as gases into the stratosphere, where they would be expected to oxidize into sulfate particles with the characteristic size of several tenths of a micron. To me, this would be a disaster. Oceans and lakes would go acidic, killing everything in them. We would be subjecting the world to increased acidification from both the sulfur as well as the existing CO_2. This method does not change

CO_2 concentrations; it simply blocks out light. (The same complaint holds for all the sun-shading ideas.)

The technological fixes outlined here are by no means the only anti–climate change proposals being suggested. Others include seeding the clouds, building ocean-cooling pipes, and changing the salinity of the sea. There is no one simple answer to tackling global warming. In most cases, one solution generates another set of problems and fears. Even the popular environmentalist solution—an injunction for us to simply "consume less"—is an economic and political minefield. If the West drastically reduces consumption, then what happens to those emerging nations whose income and jobs depend on Western appetites for oil, gas, and out-of-season green beans?

A FINAL NOTE

When people ask about the fate of humanity—as in how long we might last—most of us seem to respond that humans might have very little time left before we somehow doom ourselves to extinction. There does seem to be an awfully powerful dose of human guilt out there. But, in fact, many of us are not so pessimistic. Most "normal" mammal species last at least several million years from their speciation to their extinction, and we humans are anything but normal. The factors that drive other species into extinction—a new disease, a new predator, loss of food supply, even climate change—probably will not do us in. Perhaps only a truly colossal asteroid or comet impact on the earth can now do it. Yet survival does not mean a high quality of life. How can we avoid the reduction of living standards, the diminishment of civilization, and the lowering of the mean life expectancy that could come about if the economies of the world are struggling to build new ports, seawalls and dikes, new railroad lines, even new cities? No one now reading this book will ever know for sure whether the rise in sea level will actually occur by the end of this century as rapidly and as catastrophically as I have outlined in this book. I do not believe that China, India, or the rapidly burgeoning and technologically advancing countries such as Brazil and Indonesia will give up their new cars, let alone make all the other behavioral changes necessary, just as I do not believe that the Western countries will actually come up with the billions of dollars promised at Copenhagen to catalyze the planting of vast new forests in poorer equatorial nations. Can we actually accomplish the behavioral changes and technological feats listed above? No matter whether we believe that we will eventually succeed or fail, we must try. Yet what if we do not rise to the challenge? I end this book with two scenarios far less hopeful than the one that began this chapter. Here are two fables of what things might be like if we do not reduce emissions and fail to stop the seemingly inexorable flooding of the earth. Let it be a cautionary tale.

Bangladesh-India border, 2400 CE. CO$_2$ at 1,200 ppm.
Sea level has risen 24 feet.

The young man watched as sappers blew the barbed wire marking the border Bangladesh shared with India along the divided town of Khowai. The old, black-market tanks and armored personnel carriers, gasoline powered in this era of almost no gasoline to be had, lumbered through the wide gap and by their very presence demonstrated the point to which Bangladeshis' individual desperation overmatched the government's more common sense—and military understanding of the relative positions of India and its literally dirt-poor next-door neighbor. Yet more than half of Bangladesh was now underwater, and much of the rest might as well have been, for the cruel physics of laterally moving salt had wrecked most agricultural production. The old capital of Dacca was long gone, with only the minaret turrets and buildings of greater than three stories poking out of the water in dispiriting and rapidly rotting states. With no jobs, no food, and almost no hope, thousands had volunteered to be part of a first wave of suicide troops, their zeal shadowed only by their emaciation from lack of protein in their diet. Thus it was that on a warm May morning in stifling humidity that a tide of humanity marched right up to the great barbed-wire fence between the two benighted countries (for India too was having no end of troubles from overpopulation amid too little food in the wake of unceasing drought, but at least the Indians had lots of dry land to stand on).

More and more thousands crowded forward and were elated to find themselves pulled and pushed by the crowd into India itself. Happily, little fire came from the Indian side of the border. A few of the new arrivals had even brought passports, but these were quickly put out of sight under the ridicule of the crowd. Everyone, however, had a pack on their back and bags of possessions, for it was realized that this would be a one-way trip.

This part of the border had been carefully chosen by the Bangladeshi authorities, because it was relatively lightly guarded, and because there were many Muslims here of Indian descent. Of course, in the face of the belligerent scream soon to come from the Indian government, the worried men in Bangladesh's government would claim total ignorance of an emigration totaling millions of citizens. Yet there was no choice. Bangladesh was in the throes of disappearing from the roll of nations. While elevations in the extreme north reached over a thousand feet, most of the country was within 10 feet of sea level. This part of India could support and feed many more people than lived there now—the Indians were hoarding their land. They needed to share. That was certainly God's will.

The border between the two countries was long indeed—almost 2,500 miles long, in fact. But that entire length was a no-man's land of barbed wire and garish floodlight towers, built first in 2008 to try to stop the Bangladeshi infiltration into India. Back then, before the rising, the number of illegal border crossings was small. But as the sea rose and the population was pushed ever farther inland, it eventually backed into this fatal border region. What once had been the country with the highest number of humans per square mile was twice as packed as before.

The Indians were not much on neighborliness. In 2100 the exclaves of Dohogram-Angorpotha and Dehgram-Angalpota had been surrounded by Indian Border Security Forces (BSF, which was in reality now an army unit); the Teen Bigha Corridor, supposedly leased to Bangladesh for perpetuity as a means of accessing the exclaves from the mother country's real border, had been seized. Dohogram-Angorpotha had an area of about 10 square miles, on which 30,000 Bangladeshis tried to make a living, but in recent times they had been starving to death.

Bangladesh was once the size of Illinois, and way back in the middle of the twenty-first century "only" had a population equal to that of the entire United States at that time—more than 300 million. They had farmed to the limit of the land's ability: before the flooding of 2100 they had 8.774 million hectares of cultivable land, of which 88 percent had been cultivated, using intensive double and sometimes triple cropping on the same land to effectively increase crop production by 150 percent. But now 40 percent of Bangladesh was warm, shallow sea. The land was exhausted or salty. Famine had killed over a million Bangladeshis in the past decade, and that number was just the beginning.

In 2010 food production outpaced the needs of the ever-growing population at even the near-starvation diet of a pound of food-grains, with an average family of five eating just over 1,800 pounds annually. With average agricultural land holding at 0.2 acres and average production at 4 metric tons per acre, the typical agricultural family in 2010 was able to produce almost exactly its intake requirements. But then the land was overwhelmed by floods of sea and babies.

Many were too weak to continue through the blown fence; many were too frightened by the increasing noise as the Indian defense force began throwing mortar rounds into the Bangladeshis pouring through the broken wall. The young Bangladeshi was now through the fence and into India, trying not to look at his shattered countrymen scattered across the field, most with legs blown off from the mines. He dashed on as the first jets roared overhead, and now there was hand-to-hand fighting as the Bangladeshis reached the barracks of the Indian troops and overran them. "Run" was the operative word. This was not a conventional battle, where the object was to take an enemy stronghold, killing as many opposition soldiers as possible. This was a land grab.

Now out of breath, the Bangladeshi stopped running and surveyed the scene. The Indian troops were in full retreat. *Funny,* he thought—*they are running away from an attack on their sovereign territory.*

An hour passed, and there were now tens of thousands of Bangladeshis through the border. Many had gone back across and gathered personal possessions dropped in the rush across the border; many were now leading a valuable cow or ox into India, to find land. An Indian major watched all of this transpire from his bunker, hidden in the rugged foothills about 2 miles away. His radio squelched to life, and the major stiffened as he recognized the voice of the prime minister. The orders were succinct. "You are sure that the wind blows away from us? Yes? Then you may commence firing with the low yield first. Target the main mass of people, and then lob the higher-yield shells into the river. It only goes downstream from there, into those devils' country. Fire now."

The major steeled himself. He turned to the waiting crews. Goggles on, radiation vests on, target and fire. The modified howitzers belched flame and sound as the large shells blinked into the sky and disappeared simultaneously, hitting their high point and then letting gravity do the rest.

Before the third and fourth shells could be fired, an enormous flash of light and a roar unlike anything any of them had ever heard came from the border. The mushroom clouds from the two tactical nuclear warheads merged into one. Fifty thousand Bangladeshis died instantly. Many more would linger for a week or so. The next rounds targeted the river, vaporizing it. But the water would come back, and then it would pick up the radiation and carry it right through the heart of the Bangladeshi towns and fields. One could never be too careful about exterminating out-of-control populations.

Old Seattle, 5515 CE. Carbon dioxide at 1,200 ppm.

The small hill, only a few dozen feet above sea level, was where Anthropogenic-age Queen Ann Hill had stood, part of old Seattle. Seattle used to be composed of eight large hills, the highest 800 feet above sea level. Now they were considerably lower in elevation. Thanks to the new sea level, Seattle was now seven islands and a peninsula. To the south, where the nearby shoreline had once been, legend said that an ancient pillar known as the Space Needle had extended 300 feet out of the still water, with another 300 under the sea.

No one called it Seattle anymore. This place did not deserve a name. Sea level had risen 240 feet before it stopped. Gradually, the human race, considerably reduced in numbers, had gotten used to this new level. And then, gradually, the poison started breathing out of the sea, a fetid bad breath left over from the 4 billion– to 2 billion–year-old Archean Era of the early earth. The poison had been newly resurrected by a warming of the planet to the degree that the thermohaline circulation currents stopped; all major ocean currents went still. It took only about 2,000 years for the oceans to lose oxygen at their bottom, because once the currents were shut down, so too were the mechanisms that kept the deep ocean oxygenated. After that, it took just 500 more years for this oxygen-free water mass, filled with hydrogen sulfide–producing bacteria, to reach the photic zone—the depths above which plants can photosynthesize in the sea. Once that happened, a new wave of bacterial bloom occurred: green and purple sulfur bacteria. Once the surface of the global ocean was bereft of oxygen but replete with hydrogen sulfide and phosphate and nitrate nutrients, and all the oxygen-dependent creatures small to large had died in the transition, the surface seas had filled with the green and the much-more-visible purple sulfur microbes. The color of the ocean changed. To humans, purple became the color associated with death. No longer did widows wear black.

Any strong gust of wind from the nearby sea brought the rotten-egg stench to this onetime city. The seaway itself, part of old Puget Sound, was made up of smooth, wave-free water, clear and beautiful under the hot tropical sun, fringed by palms and other tropical plants. The entire vast waterway lapped right onto the Olympic Mountains to the west, which now were islands. The colors of the sea were spectacular in the bright sun, perhaps made all the more vivid because of the high ultraviolet content of the sunlight. The shallows with their microbiolite reefs showed the green, red, and brown colors characteristic of such reefs— which were now virtually global in extent. Farther out the ocean was a deep purple, painted that vivid hue by the untold billions of purple sulfur bacteria that lived in the sun-dappled upper parts of the sea—a place that, like the bottom, held no oxygen.

To a diver from the twentieth century, or even from the first few decades of the twenty-first, the reefs would look strange indeed. Coral reefs throughout time have been characterized by their high diversity of life, and some of the most spectacularly diverse groups were always the small reef fish. Millions of them would dance in the sun, busily going about their business amid the tropical coral reefs. But corals were now mostly gone, and they were not part of these reefs; no coral could live in the 90-degree, largely anoxic body of water that Puget Sound had become. The bottoms of that sea were sterile of all but microbes; only a thin slick of oxygenated water could be found. Nevertheless, once in a while a tropical mackerel or wahoo would stray from the open sea, where there were still regions of oxygenated surface water. The Pacific was faring better than the Atlantic in terms of maintaining its oxygenated deep-bottom waters. The rise of anoxia was neither worldwide nor evenly distributed. But the many fjords newly formed along the Pacific Coast had quickly gone eutrophic and oxygen free. Western Oregon was a large inland sea, with highlands still rising out of the water to its west, but that giant new seaway was home only to microbes, many busily engaged in producing hydrogen sulfide, and in this way was similar to all of the traditionally quieter waters and smaller seas, places like the new Gulf of Mexico, which now extended well into Missouri and parts of the dust-bowl desert formerly known as Oklahoma; the entire Mediterranean Sea; and Hudson Bay.

Puget Sound and the Straits of Georgia were among the first to go largely anoxic, soon after the completion of the 240-foot rise in sea level. Salmon had disappeared early in the twenty-first century, and while fish farming had kept the stocks alive for a while, eventually the tropical conditions and flooding of the rivers finished off both wild and farm-raised fish, thanks to the steadily rising chemocline—the border between oxygenated surface water and anoxic bottom water.

Oil, followed by coal, had gradually run out. By necessity, humanity reverted to a low-tech, pastoral lifestyle. Infant mortality increased; old age became rare. Because there was no coal, there could be no smelting of metals, and without petroleum, there was no more plastic. Human population worldwide had subsided to less than a billion people, living largely on yams, taro, tapioca, potatoes, and corn. Cows, pigs, and sheep were now all extinct. The rising hydrogen sulfide content in the atmosphere had killed them off. It was often considered a happy miracle that humans could withstand higher concentrations of hydrogen sulfide than could livestock. In fact, the who's-who of survivors among mammals had little rhyme or reason. All the reptiles and amphibians thrived. Anything with cold blood thrived. But the warm bloods were hit hard.

Hardest hit of all were the birds. Their exquisite, double-pump lung system, with its intricate set of air sacs placed cleverly into the hollows of their very bones, served them well at higher altitudes and in their strenuous exercise of flying. But their very efficiency at extracting oxygen molecules also brought in far more hydrogen sulfide than what any other kind of animal internalized. While the sky was still filled with flyers, all were insects of one kind or another. All birds were now extinct.

For the surviving humans, food was always a problem. Long gone were the "megafarms" of previous millennia, because there was no fuel for tractors, and then no tractors. There was no electricity because metallurgy had just about ceased. The only metals were those scavenged from the wastes left behind by the pre-Flood people, but each year oxidation and rust returned the metals to more oxidized minerals. Metal did not last long even in the relatively mild monsoon season of the Seattle region, the six months when it rained only 200 inches and the temperature rarely exceeded 110 degrees.

It had been a powerful thing, that global warming and the truly apocalyptic sea level rise that it unleashed. It was the ultimate "kill" mechanism between a warmed planet and its dying animals and plants.

NOTES

INTRODUCTION: MIAMI BEACHED

1. Description of the amount and areas of flooding that Miami will experience under various sea level rises is found at South Florida Geographic Information Systems, www.sfrpc.com/gis/slr.htm, in a PDF of a Miami–Dade County map.

2. Sea level rise from tectonic forces is described in any number of college-level texts on stratigraphy, such as Steven Stanley, *Earth System History* (New York: W. H. Freeman, 1999).

3. Many articles profile this kind of sea level change. Two good ones are R. B. Alley, P. U. Clark, P. Huybrechts, and I. Joughin, "Ice-Sheet and Sea-Level Changes," *Science* 310, no. 5747 (October 21, 2005): 456–460; and R. G. Fairbanks, "A 17,000 Year Glacio-Eustatic Sea Level Record: Influence of Glacial Melting Rates on the Younger Dryas Event and Deep Ocean Circulation," *Nature* 342 (December 7, 1989): 637–642.

4. The history of sea level in North America can be found in Stanley, *Earth System History*, while discussions about the Cretaceous Tertiary extinction can be found in Peter Ward, *Rivers in Time* (New York: Columbia University Press, 2002).

5. Pleistocene sea-level history of the Florida Keys is from B. Lidz, "Pleistocene Corals of the Florida Keys: Architects of Imposing Reefs—Why?" *Journal of Coastal Research* 22, no. 4 (July 2006): 750–759.

CHAPTER 1: THE RISING SEA

1. We were stationed at James Ross Island, Antarctica, from mid-February to mid-March 2009 and mid-November to late December 2009.

2. *Science* 319, no. 5860 (January 11, 2008): 189–192.

3. The change of paradigm for mass extinctions comes from numerous sources, summarized in my October 2006 *Scientific American* article, my book *Under a Green Sky*

(Washington, DC: Smithsonian Books, 2007), and many scientific articles, including R. B. Huey and P. D. Ward, "Hypoxia, Global Warming, and Terrestrial Late Permian Extinctions," *Science* 308, no. 5720 (March 15, 2005): 398–401; K. Farley, P. Ward, G. Garrison, and S. Mukhopadhyay, "Absence of Extraterrestrial 3He in Permian Triassic Rocks," *Earth and Planetary Science Letters* 240 (2005): 265–275; and P. D. Ward, J. Botha, R. Buick, M. O. De Kock, D. H. Erwin, G. Garrison, J. Kirschvink, and R. M. H. Smith, "Abrupt and Gradual Extinction Among Late Permian Land Vertebrates in the Karoo Basin, South Africa," *Science* 307, no. 5710 (February 5, 2005): 709–714.

4. See my October 2006 *Scientific American* article expanding on this.

5. See www.smm.org/deadzone for a good description of these terrible dead zones.

6. The story of the dead French horse can be found in the AP news release at www.nola.com/news/index.ssf/2009/08/toxic_beach_issue_gains_promin.html.

7. The name "greenhouse extinctions" has been informally used, and I first began using it in a 1996 *Discover* magazine article titled "Greenhouse Extinctions." It probably was used long before this.

8. His paper can be found in *Nature* 457 (January 22, 2009): 459–462.

9. According to many sources, such as the NPR Web site story of May 15, 2009, www.npr.org/templates/story/story.php?storyId=104133389.

10. See http://home.att.net/~atrahasis/index.htm.

11. They recounted this evidence in their book *Noah's Flood: The New Scientific Discoveries about the Event that Changed History* (New York: Simon & Schuster, 1998).

12. See http://kingfish.coastal.edu/biology/sgilman/778AnimalAdapt.htm for more on this adaptation.

13. See www.pbs.org/saf/1207/features/noah.htm.

14. H. E. Wright, "Global Climatic Changes Since the Last Glacial Maximum: Evidence from Paleolimnology and Paleoclimate Modeling," *Journal of Paleolimnology* 15, no. 2 (March 1996).

15. *Climate Change 2007*, the Fourth Assessment Report of the IPCC. For a clear explanation see the assessment at www.realclimate.org/index.php/archives/2007/03/the-ipcc-sea-level-numbers.

16. One such article is from the *Christian Science Monitor*: http://features.csmonitor.com/environment/2009/05/14/new-study-less-sea-rise-expected-from-possible-antarctic-melt/.

17. This can be found at www.ipcc.ch/pdf/assessment-report/ar4/wg2/ar4-wg2-spm.pdf.

18. An example is the 4th International Conference on Integrating GIS and Environmental Modeling (GIS/EM4): "Problems, Prospects and Research Needs," Banff, Alberta, Canada, September 2–8, 2000: GIS-based modeling of sea-level rise impacts for coastal management in southeastern Australia.

19. *Climate Change 2007: The Physical Basis*, 4th Assessment Report, IPCC, ed. S. D. Solomon and Q. M. Manning (Cambridge: Cambridge University Press, 2007).

20. Robert J. Nicholls and Richard S. J. Tol, "Impacts and Responses to Sea-Level Rise: A Global Analysis of the SRES Scenarios over the Twenty-First Century," *Philosophical Transactions of the Royal Society A* 64, no. 1841 (April 15, 2006): 1073–1095.

21. Neville Nicholls, "Estimating Changes in Mortality Due to Climate Change," *Climatic Change* 97 (2009): 1–2; N. Nicholls, "Long-Term Climate Monitoring and Extreme Events," *Climatic Change* 31 (1995): 231–245.

22. A good source is www.climate.org/topics/sea-level/index.html.

23. While I promised myself I would never do it, the Wikipedia story of Hurricane Katrina is as good as any: http://en.wikipedia.org/wiki/Hurricane_Katrina.

24. For an eye opener, look at the official tide changes: www.sailwx.info/tides/fundy.phtml.

25. His column is at www.washingtonpost.com/wp-dyn/content/article/2009/04/01/AR2009040103042.html.

26. The history of the controversy is found in many sources, including my own *Under a Green Sky*.

27. Hansen, "Scientific Reticence and Sea Level Rise," Environmental Research Letters 2 (April–June 2007), www.iop.org/EJ/article/1748–9326/2/2/024002/erl7_2_024002.html.

28. Stefan Rahmstorf, "A Semi-Empirical Approach to Projecting Future Sea-Level Rise," *Science* 315 (2007): 368–370; see also www.guardian.co.uk/environment/cif-green/2009/mar/03/sea-levels-rising. Previous articles on this topic include: S. Rahmstorf, "Shifting Seas in the Greenhouse?" *Nature* 399 (June 10, 1999): 523; S. Rahmstorf, "Ocean Circulation and Climate During the Past 120,000 Years," *Nature* 419 (September 12, 2002): 207; S. Rahmstorf and A. Ganopolski, "Long-Term Global Warming Scenarios Computed with an Efficient Coupled Climate Model," *Climatic Change* 43, no. 2 (October 1999): 353.

CHAPTER 2: RISING CARBON DIOXIDE

1. Comments on tar sands in general can be found at www.guardiuk/an.co.environment/cif-green/2009/mar/03/sea-levels-rising. For the Athabasca case, and whether it can be cleaned up, go to www.newscientist.com/article/mg20227043.900-can-oil-from-tar-sands-be-cleaned-up.html, from *New Scientist*. The effects on the Canadian indigenous people can be found at www.indiancountrytoday.com/global/40856617.html with the article "Tar Sands Are Killing Us," by Kate Harries, *Indian Country Today* correspondent, March 11, 2009. A good scientific source on pollution is K. P. Timoney and P. Lee, 2009, "Does the Alberta Tar Sands Industry Pollute? The Scientific Evidence," *Open Conservation Biology Journal* 3 (2009): 65–81.

2. In a new study, the British Meteorological Department has warned that global warming could result in a rise of 7 degrees Fahrenheit by the year 2060. The report can be obtained by contacting the British Department of Energy and Climate Change, United Kingdom's Met Hadley Centre for Climate Prediction and Research.

3. Don Brownlee and I discuss the formation of elements in our book *The Life and Death of Planet Earth* (New York: Henry Holt, 2003). This book also sources the rest of this paragraph.

4. A good source on the discovery of carbon dioxide is at http://en.wikipedia.org/wiki/Carbon_dioxide. A fuller treatment can be found in Mark Zachary Jacobson, *Atmospheric Pollution: History, Science, and Regulation* (Cambridge: Cambridge University Press, 2002).

5. An excellent summary of the history of greenhouses can be found at www.hobby-greenhouse.com/history_of_greenhouses.htm.

6. This wondrous edifice nudged me early on to spend my life studying paleontology—not for the plants, but for the first reconstructions of dinosaurs, the genus *Iguanodon* to

be exact. It was portrayed as a particularly ugly frog. A wonderful, illustrated site on this is www.victorianstation.com/palace.html.

7. See the Nobel Web site for a biography of Arrhenius: http://nobelprize.org/nobel_prizes/chemistry/laureates/1903/arrhenius-bio.html.

8. This and many more interesting facts about carbon dioxide and its cycles can be found at www.waterencyclopedia.com/Bi-Ca/Carbon-Dioxide-in-the-Ocean-and-Atmosphere.html. A good academic source is the wonderful book by my friend and colleague Dr. Robert Berner of Yale University, *The Phanerozoic Carbon Cycle* (Cambridge: Cambridge University Press, 2003).

9. Roger Revelle and Hans E. Suess, "Carbon Dioxide Exchange Between Atmosphere and Ocean and the Question of an Increase of Atmospheric CO_2 During the Past Decades," *Tellus* 9 (1957): 18–27. See a fuller account of this paper and the discovery of the greenhouse effect at www.aip.org/history/climate/Revelle.htm.

10. The official NOAA Web site is incredibly useful about climate change. Not only this particular issue but also most important points about the relationship between CO_2 and temperature can be found at www.ncdc.noaa.gov/paleo/globalwarming/temperature-change.html. Of course, if you want the anti–global warming perspective, a nice slanted view can be found at http://scienceandpublicpolicy.org/monckton/temperature_and_co2_change_briefing.html.

11. A history of CO_2 in the atmosphere is (once again) Roger Berner, *The Phanerozoic Carbon Cycle* (Cambridge: Cambridge University Press, 2003). I have also written two books that touch on the subject: *Under a Green Sky* (New York: HarperCollins, 2007), and *Out of Thin Air* (Washington, DC: Joseph Henry Press of the National Academy of Sciences, 2006). Other articles include R. B. Alley, P. U. Clark, P. Huybrechts, and I. Joughin, "Ice-Sheet and Sea-Level Changes," *Science* 310, no. 5747 (October 21, 2005): 456–460.

12. Articles about this are K. Caldeira and M. R. Rampino, "The Mid-Cretaceous Superplume, Carbon Dioxide, and Global Warming," *Geophysical Research Letters* 18, no. 6 (1991): 987–990; E. J. Barron, P. J. Fawcett, W. H. Peterson, D. Pollard, and S. L. Thompson, "A 'Simulation' of Mid-Cretaceous Climate," *Paleoceanography* 10, no. 5 (1995): 953–962; E. J. Barron, W. W. Hay, and S. Thompson, "The Hydrologic Cycle, a Major Variable During Earth History," *Palaeogeography, Palaeoclimatology, Palaeoecology* (Global and Planetary Change section) 75, no. 3 (1989): 157–174.

13. The best paper about this comes from one of my past students, Brian T. Huber, Richard D. Norris, and Kenneth G. MacLeod, "Deep-Sea Paleotemperature Record of Extreme Warmth During the Cretaceous," *Geology* 30, no. 2 (2002): 123–126.

14. Ice core work (including at Lake Vostok) is reviewed in M. J. Siegert, *Ice Sheets and Late Quaternary Environmental Change* (Hoboken, NJ: Wiley & Sons, 2001).

15. CO_2 levels with time: Chapter 1 and references above, as well as Hubertus Fischer, Martin Wahlen, Jesse Smith, Derek Mastroianni, and Bruce Deck, "Ice Core Records of Atmospheric CO_2 Around the Last Three Glacial Terminations," *Science* 283, no 5408 (March 12, 1999): 1714–1717. For more modern CO_2 levels see the instructive NOAA Web site, www.ncdc.noaa.gov/paleo/ctl/cliihis100.html.

16. A short (102 pages) but readable primer on ocean chemistry as it relates to the topic mentioned here is R. E. Hester and R. M. Harrison, eds., *Chemistry in the Marine Environment* (Cambridge: Cambridge University Press, 2000).

17. There are many explanations of how orbital position affects climate. The site www.windows.ucar.edu/tour/link=/earth/climate/cli_sun.html is well illustrated and to the

point. Another good site is from the U.S. Navy: http://aa.usno.navy.mil/faq/docs/seasons_orbit.php.

18. The best commentary on this comes in a column from Dave Archer, at www.realclimate.org/index.php/archives/2005/03/how-long-will-global-warming-last.

19. www.realclimate.org/index.php/archives/2006/01.

20. Hansen is a seminal figure and avowed lightning rod about climate change. His papers on the subject include Hansen, et al., "Global Warming in the Twenty-First Century: An Alternative Scenario," *Proceedings of the National Academy of Sciences* 97, no. 18 (August 29, 2000): 9875–9880; Hansen, et al., "Climate Change and Trace Gases," *Philosophical Transactions of the Royal Society* A 365, no. 1856 (July 15, 2007): 1925–1954.

It also appears that Hansen's claims that were censored by officials in the George W. Bush administration can be found in Andrew C. Revkin, "Climate Expert Says NASA Tried to Silence Him," *New York Times*, January 29, 2006, www.nytimes.com/2006/01/29/science/earth/29climate.html. See also Andrew C. Revkin, "NASA Office Is Criticized on Climate Reports," *New York Times*, June 3, 2008, www.nytimes.com/2008/06/03/science/earth/03nasa.html.

21. The yearly change in CO_2 is shown by NOAA, www.esrl.noaa.gov/gmd/ccgg/trends/.

22. http://news.bostonherald.com/news/opinion/op_ed/view.bg?articleid=1210332&srvc=home&position=emailed. Also, see commentary in *Nature* 449 (October 25, 2007): 973–975, published online October 24, 2007.

23. Estimated world carbon demands are from a variety of sources, including http://afp.google.com/article/ALeqM5gPGhbOgx6LbwIQhLcUmINFhlGCZw, and this report from the International Energy Agency, www.iea.org/press/pressdetail.asp?press_rel_id=187; also, a pretty cool interactive Web site is www.breathingearth.net, and this illustrates the basic problem. However, when I see sites like these (and the analogous sites where one can input sea level rise figures for a given geographic area, and watch how quickly it melts), they feel like video games. The message seems lost.

24. The concept of a climate tipping point can be found at www.time.com/time/health/article/0,8599,1920168,00.html (and many other sites) including www.motherjones.com/environment/2006/11/thirteenth-tipping-point. The always reliable *New Scientist* also commented as follows: www.newscientist.com/article/dn17680-climate-tipping-point-defined-for-us-crop-yields.html.

25. The Hansen et al. article quoted here is www.columbia.edu/~jeh1/2008/TargetCO2_20080407.pdf.

26. Ibid.

CHAPTER 3: THE FLOOD OF HUMANS

1. My own visits to Tunisia and the area were in 1992 and 2000. At that time I was astonished at the number of children as a visible proportion of the population.

2. David Archer, *The Long Thaw: How Humans Are Changing the Next 100,000 Years of Earth's Climate* (Princeton, NJ: Princeton University Press, 2009).

3. This well-funded organization later changed its name to Population Connection (www.envirolink.org/resource.html?itemid=785&catid=5). In the long term, zero population growth is achieved when a population's birthrate equals its death rate.

4. The many population estimates in this chapter come from the United Nations. The most useful is from a 2004 publication put out by the Department of Economic and Social

Affairs, Population Division, titled *World Population to 2300*. Another useful resource is the UN site: http://unstats.un.org/unsd/mdg/Metadata.aspx?IndicatorId=0&SeriesId=566.

5. The source for this is the World Factbook, produced regularly by the CIA and found at https://www.cia.gov/library/publications/the-world-factbook/geos/vt.html.

6. Once again, refer to the 2004 publication from the UN Department of Economic and Social Affairs, Population Division, *World Population to 2300*.

7. EIA, *International Energy Outlook*, DOE/EIA-0484(2009). Instructions for obtaining the report can be found at www.eia.doe.gov/oiaf/ieo.

8. Coal as the leading source of future global energy: http://news.alibaba.com/article/detail/business-in-china/100184090-1-china-coal-energy%2527s-raw-coal.html, and the peer-reviewed M. Höök and K. Aleklett, "Historical Trends in American Coal Production and a Possible Future Outlook," *International Journal of Coal Geology* 78, no. 3 (2009): 201–216.

9. The figures for the following paragraphs, including the definition of Hubbert's peak as well as oil production trends, reserves, and current oil fields, can be found at www.hubbertpeak.com/ and K. Deffeyes, *Beyond Oil: The View from Hubbert's Peak* (New York: Farrar, Straus and Giroux, 2005).

10. Oil consumption comes from a 2009 *Wall Street Journal* article, http://online.wsj.com/article/SB125801667159145091.html?mod=WSJ_hpp_MIDDLENextto WhatsNewsTop, as well as from periodic reports from the International Energy Agency, a non-political organization not associated with any single country. Its Web site is www.iea.org, and the relevant report is the *World Energy Outlook 2009*.

11. The 2005 Hirsch Report, formally titled *Peaking of World Oil Production: Impacts, Mitigation, and Risk Management*, is published by the U.S. Department of Energy and is available at www.nyswda.org/LegPosition/HirschReport.htm, as is *The Inevitable Peaking of World Oil Production*, a shorter version of the Hirsch Report.

12. Types of coal: www.appaltree.net/aba/coaltypes.htm.

13. Ibid.

14. A summary of the Obama efforts for "clean coal plants" is at www.usnews.com/articles/news/energy/2009/06/12/obama-administration-pouring-1-billion-into-clean-coal-project.html. However, skepticism that such clean coal plants are possible can be found at *Scientific American*, July 21, 2009, www.scientificamerican.com/article.cfm?id=obama-and-clean-coal.

15. How to "clean" coal can be found in this article: Sarah Dowdey, "What Is Clean Coal Technology?" HowStuffWorks, http://science.howstuffworks.com/clean-coal.htm.

16. The expense of clean coal technology is more than the Chinese government wants to bear: www.watoday.com.au/environment/climate-change/china-faces-massive-bill-for-clean-coal-20090905-fbpm.html.

17. David Biello, "China Targets Cleaner–Coal Power Plants," *Scientific American*, October 2008, www.scientificamerican.com/article.cfm?id=china-targets-cleaner-coal.

18. A concise and useful description of the GreenGen coal plant is found in the above Biello article in *Scientific American*. Other articles in the October 2008 special issue are also very useful.

19. This figure comes from the report *Food, Land, Population and the U.S. Economy* by David Pimentel and Mario Giampietro. The full report can be obtained from Carrying Capacity Network, 2000 P Street NW, Suite 240, Washington, DC 20036. Also, a useful site is www.futurepundit.com/archives/005135.html.

20. The 2008 UN report on global food prices can be found at www.un.org/millenniumgoals.

21. Information about Goodchild's claims about human food and mortality can be found in the Carbon Report, www.carboncapturereport.org/cgi-bin/biodb?PROJID=3&mode=viewpersonname&name=peter_goodchild. One of his books is *Survival Skills of the North American Indians*.

22. Estimates on the number of cars can be found at www.123helpme.com/view.asp?id=23459.

23. Glenn Elert, ed. *Physics Factbook*, http://hypertextbook.com/facts/2001/Marina Stasenko.shtml.

24. A good source for the relative importance of various carbon emissions going into the atmosphere is the U.S. Environmental Protection Agency, www.epa.gov/OMS/climate/420f05004.htm.

25. From the National Defense Research Council, www.nrdc.org/globalWarming/f101.asp.

26. Specific numbers of cars owned by Americans and Chinese can be found at www.peopleandplanet.net/doc.php?id=2484.

27. A good review (well, actually it is reviewed as being bad) can be found at the National Public Radio report www.npr.org/templates/story/story.php?storyId=17984516.

28. A series of graphs showing the relative carbon emissions by country is at http://rainforests.mongabay.com/09-carbon_emissions.htm.

CHAPTER 4: FEEDING HUMANITY AMID RISING SEA LEVEL

1. A report on this can be found in Z. Su et al., "Drought Monitoring and Prediction over China," Proc. Dragon 1 Programme Final Results 2004–2007, April 21–25, 2008 (ESA SP-655, April 2008).

2. The Chinese Web site about the Three Gorges Dam is www.ctgpc.com. However, a far less glowing report about it was produced by the International Rivers organization: www.internationalrivers.org/en/china/three-gorges-dam.

3. The start of human agriculture and subsequent history is usefully discussed by Jared Diamond (author of *Guns, Germs, and Steel*, among other books) at www.unl.edu/rhames/courses/orig_agri_tur.html.

4. Figures come from the USDA, which puts out myriad reports each year on its Web site. The complete 2008 edition is at www.nass.usda.gov/Newsroom/2008/08_12_2008.asp.

5. A scathing report on why the Green Revolution in agriculture is going wrong is at www.commondreams.org/view/2009/10/05-9.

6. This article explains part of the picture about plant yield and CO_2: P. J. Gregory, L. P. Simmonds, G. P. Warren, T. Batey, J. K. Syers, J. S. Wallace, and M. V. K. Sivakumar, "Interactions Between Plant Nutrients, Water and Carbon Dioxide as Factors Limiting Crop Yields [and Discussion]," *Philosophical Transactions of the Royal Society B* 352, no. 1356 (July 29, 1997): 987–996. Also an interesting paper from physicians' point of view is by A. Robinson, N. Robinson, and W. Soon, "Environmental Effects of Increased Atmospheric Carbon Dioxide," *Journal of American Physicians and Surgeons* 12, no. 3 (Fall 2007): 79–90.

7. D. S. Battisti and R. L. Naylor, "Historical Warnings of Future Food Insecurity with Unprecedented Seasonal Heat," *Science* 323, no 5911 (January 9, 2009): 240–244. A further report is Stephen P. Long, Elizabeth A. Ainsworth, Andrew D. B. Leakey, Josef

Nösberger, and Donald R. Ort, "Food for Thought: Lower-Than-Expected Crop Yield Stimulation with Rising CO_2 Concentrations," *Science* 312, no. 5782 (June 30, 2006): 1918–1921, and references therein.

8. William R. Cline, *Global Warming and Agriculture: Impact Estimates by Country* (Washington, DC: Peterson Institute for International Economics, 2007).

9. William R. Cline, *The Economics of Global Warming* (Washington, DC: Peterson Institute for International Economics, 1992).

10. The FAO perspective can be found in the report "World Agriculture: Towards 2015/2030" at www.fao.org/docrep/005/y4252e/y4252e12a.htm.

11. The lag effect can be found in P. Ciais, P. P. Tans, M. Trolier, J. W. C. White, and R. J. Francey, "A Large Northern Hemisphere Terrestrial CO_2 Sink Indicated by the $^{13}C/^{12}C$ Ratio of Atmospheric Carbon Dioxide," *Science* 269, no. 5227 (August 25, 1995): 1098–1102.

12. Among the many good explanations of this is a Duke University report: www .biology.duke.edu/jackson/post1.html.

13. A nice explanation of C_4 versus C_3 plants is at www.answers.com/topic/c3-and -c4-plants.

14. For the real "Bible" about C_3 and C_4 plants, see T. E. Cerling and Maria-Denise Dearing, *A History of Atmospheric CO2 and Its Effects on Plants, Animals, and Ecosystems* (New York: Springer, 2005).

15. For more on C_3 plants of agricultural importance: J. C. Forbes and R. Drennan Watson, *Plants in Agriculture* (Cambridge: Cambridge University Press, 1992).

16. Sylvan Harold Wittwer, *Food, Climate, and Carbon Dioxide: The Global Environment and World Food* (New York: CRC Press, 1995).

17. See the October 1, 2009, *Economist* article on this at www.economist.com/world/ international/displaystory.cfm?story_id=14540051.

18. Michael Timberlake and Jeffrey Kentor, "Economic Dependence, Overurbanization, and Economic Growth: A Study of Less Developed Countries," *Sociological Quarterly* 24, no. 4 (Autumn 1983): 489–507.

19. The 2050 Project looks at future crop yields in various countries by 2050, and the news is not good: http://the2050project.com/index.php?option=com_content&task=view &id=75.

20. Robert Mendelsohn, Michael Schlesinger, and Larry Williams, "Comparing Impacts across Climate Models," *Integrated Assessment* 1, no. 1 (March 2000): 37–48.

21. The very useful Chapter 4, "Country Level Climate Projections," in *Global Warming and Agriculture, Impact Assessments by Country*, can be downloaded in PDF form at www.cgdev.org/doc/books/Cline%20global%20warming/Chapter%204.pdf.

22. IPCC 4, referenced above.

23. Environment International.

24. For more on precipitation pattern, see the following article: M. Dore, "Climate Change and Changes in Global Precipitation Patterns: What Do We Know?" *Environment International* 31, no. 8 (October 2005): 1167–1181.

25. D. Battisti and R. Nalor, "Historical Warnings of Future Food Insecurity with Unprecedented Seasonal Heat," *Science* 323, no. 5911 (January 9, 2009): 240–244, and Robert Mendelsohn and James E. Neumann, eds., *The Impact of Climate Change on the United States Economy* (Cambridge: Cambridge University Press, 2004).

26. Climate Change 2007, the IPCC Fourth Assessment Report.

27. For more information on this model, see www.sciencemag.org/cgi/content/full/312/5782/1918.

28. For those who wish to delve more deeply into this area, a PDF of more than 150 references dealing with the topic discussed here can be found at www.cgdev.org.

29. *New Scientist* covers the Swiss study at www.newscientist.com/article/mg20026803.600-hunger-hotspots-of-the-future-revealed.html.

30. A highly readable history of the Sacramento Delta region and its development is at www.sacdelta.com/hist.html.

31. A thorough and exhaustive (as well as exhausting to read) book on California water is Norris Hundley Jr., *The Great Thirst, Californians and Water: A History*, rev. ed. (Berkeley: University of California Press, 2001).

32. A good source for both maps and future projects can be found at the Web site of the Delta Initiative, a University of California at Berkeley study: http://landscape.ced .berkeley.edu/~delta.

33. Renee Montagne, "Crucial California Delta Faces a Salty Future," National Public Radio, January 14, 2008, www.npr.org/templates/transcript/transcript.php?storyId =18031391. See also Melissa Block, "California Turns to Holland for Flood Expertise," NPR, January 14, 2008, www.npr.org/templates/transcript/transcript.php?storyId =18080442.

34. The Bureau of Reclamation site is www.usbr.gov.

35. Effect of drought on plants: http://ag.arizona.edu/pubs/natresources/az1136.pdf.

36. Source: Agdex 518–17, November 2001.

37. Salt use in Chicago in 2004, found at http://chicagowildernessmag.org/issues/winter2004/salt.html.

38. http://chicagowildernessmag.org/issues/winter2004/salt.html.

39. http://chicagowildernessmag.org/issues/winter2004/salt.html.

40. The effects of salt on amphibian skin can be found at http://linkinghub.elsevier.com /retrieve/pii/S0013935108002168.

CHAPTER 5: GREENLAND, ANTARCTICA, AND SEA LEVEL

1. The WAIS is of major concern for "collapse." See www.co2science.org/subject/w/summaries/waiscollapse.php for a summary. Additional information is at R. B. Alley, P. U. Clark, P. Huybrechts, and I. Joughin, "Ice-Sheet and Sea-Level Changes," *Science* 310, no. 5747 (October 21, 2005): 456–460, and J. Hansen, et al., "Dangerous Human-Made Interferences with Climate: A GISS ModelE study," *Atmospheric Chemistry and Physics* 7 (2007): 2287–2312.

2. Climatologist Eric Steig referred me to the following article, with the comment that I should read this only if "you want to have the shit scared out of you": David Pollard and Robert M. DeConto, "Modelling West Antarctic Ice Sheet Growth and Collapse through the Past Five Million Years," *Nature* 458 (March 19, 2009): 329–332.

3. J. K. Ridley, P. Huybrechts, J. M. Gregory, and J. A. Lowe, "Elimination of the Greenland Ice Sheet in a High CO_2 Climate," *Journal of Climate* 18, no. 17 (2005): 3409–3427.

4. See www.greenfacts.org/glossary/ghi/ice-cap-ice-sheet-polar-ice-cap.htm for definitions of the two terms.

5. Ibid.

6. I recommend Matthew M. Bennett and Neil F. Glasser, eds., *Glacial Geology: Ice Sheets and Landforms* (New York: Wiley, 2009), for an up-to-date discussion on these figures. Also good is William F. Ruddiman, *Earth's Climate: Past and Future* (Freeman, 2007).

7. A good report on this is Andrew C. Revkin, "Greenland Losing Ice, With or Without Lubrication," *New York Times* Dot Earth, July 14, 2008, http://dotearth.blogs.nytimes.com/2008/07/14/greenland-losing-ice-with-or-without-lubrication and Michael Reilly, "Greenland Ice Sheet Slams the Brakes On," *New Scientist*, July 3, 2008, www.newscientist.com/article/dn14251-greenland-ice-sheet-slams-the-brakes-on.html.

8. For more information on the climate, visit www.greenland.com/content/english/tourist.

9. An Inuit perspective on being colonized: http://arcticcircle.uconn.edu/HistoryCulture/petersen.html.

10. One helpful report is Seth Borenstein, "NASA Data: Greenland, Antarctic Ice Melt Worsening," *Huffington Post*, September 23, 2009, www.huffingtonpost.com/wires/2009/09/23/nasa-data-greenland-antar_0_ws_296332.html.

11. Ibid.

12. www.realclimate.org/index.php/archives/2008/09/how-much-will-sea-level-rise.

13. "Greenland's Glaciers Losing Ice Faster This Year than Last Year," Agence France-Presse, December 23, 2008, www.climateark.org/shared/reader/welcome.aspx?linkid=113954.

14. Jonathan Amos, "Greenland Glacier Races to Ocean," BBC News, December 8, 2005, http://news.bbc.co.uk/2/hi/science/nature/4508964.stm.

15. Noelle d'Estries, "Bad News: Greenland Ice Sheet Melting at Record Rate," Planetsave.com, December 11, 2007, http://planetsave.com/blog/2007/12/11/bad-news-greenland-ice-sheet-melting-at-record-rate.

16. The NASA report documenting this can be found at www.nasa.gov/vision/earth/lookingatearth/jakobshavn.html. See also Amanda Leigh Haag, "Greenland's Ancient Analogue," *Nature Reports Climate Change*, September 2, 2008, www.nature.com/climate/2008/0809/full/climate.2008.88.html.

17. "Antarctic Ice Shelf Disintegration Underscores a Warming World," National Snow and Ice Data Center press release, March 25, 2008, http://nsidc.org/news/press/20080325_Wilkins.html.

18. One such new model is reported on in "New Model Predicts a Glacier's Life," *Science Daily*, October 31, 2008, www.sciencedaily.com/releases/2008/10/081029104258.htm.

19. Richard Harris, "Predicted Impact from Antarctic Ice Melt Lessened," National Public Radio, May 15, 2009, www.npr.org/templates/story/story.php?storyId=104133389.

20. "Arctic Summer Ice Could Disappear within Decades," *Mail & Guardian Online*, October 15, 2009, www.mg.co.za/article/2009-10-15-arctic-summer-ice-could-disappear-within-decades.

21. Jesse Byock, *Medieval Iceland: Society, Sagas and Power* (Berkeley: University of California Press, 1988); Guðmundur Hálfdanarson, *Historical Dictionary of Iceland* (Scarecrow Press, 1997); and Gunnar Karlsson, *History of Iceland* (Minneapolis: University of Minnesota Press, 2000).

22. Greenland's own spin on this can be found at www.greenlandexpo.com/content/us/trade_in_greenland/mineral_resources_and_oil_exploration_in_greenland, and

for anyone who thought I was making up the story about Russians on Greenland that begins that chapter (well, I was—sort of), see Luke Harding, "Kremlin Lays Claim to Huge Chunk of Oil-Rich North Pole," *The Guardian*, June 28, 2007, www.guardian.co.uk/world/2007/jun/28/russia.oil.

23. http://pubs.usgs.gov/fs/fs2–00.

24. http://wattsupwiththat.com/2009/10/19/west-antarctic-ice-sheet-may-not-be-losing-ice-as-fast-as-once-thought-grace-readings-overestimated.

25. Minturn Wright, "The Ownership of Antarctica, Its Living and Mineral Resources," *Journal of Law and the Environment* 4 (1987); Jennifer Frakes, "The Common Heritage of Mankind Principle and the Deep Seabed, Outer Space, and Antarctica: Will Developed and Developing Nations Reach a Compromise?" *Wisconsin International Law Journal* 21 (2003): 409.

26. An article by Ivany et al. dealing with the oldest ice on Antarctica can be found at http://geology.gsapubs.org/content/34/5/377.full.

27. J. P. Kennett, R. E. Houtz, P. B. Andrews, A. R. Edwards, V. A. Gostin, M. Hajos, M. A. Hampton, D. G. Jenkins, S. V. Margolis, A. T. Ovenshine, and K. Perch-Nielsen, "Development of the Circum-Antarctic Current," *Science* 186, no. 4159 (October 11, 1974): 144–147.

28. N. J. Shackleton and J. P. Kennett, "Paleotemperature History of the Cenozoic and the Initiation of Antarctic Glaciation: Oxygen and Carbon Isotope Analyses in DSDP Sites 277, 279, and 281," in J. P. Kennett, R. E. Houtz, et al., *Initial Reports of the Deep Sea Drilling Project* 29 (Washington, DC: U.S. Government Printing Office, 1975), pp. 743–755.

29. Kennett et al., "Development of the Circum-Antarctic Current."

30. Ibid.

31. Shackleton and Kennett, "Paleotemperature History of the Cenozoic and the Initiation of Antarctic Glaciation."

32. Robert M. DeConto and David Pollard, "Rapid Cenozoic Glaciation of Antarctica Induced by Declining Atmospheric CO_2," *Nature* 421 (January 16, 2003): 245–249.

33. C. Riedl, H. Rott, and W. Rack, "Recent Variations of Larsen Ice Shelf, Antarctic Peninsula, Observed by Envisat," Proceedings of the 2004 Envisat & ERS Symposium, Salzburg, Austria, http://earth.esa.int/workshops/salzburg04/; Steve Connor, "Ice Shelf Collapse Was Biggest for 10,000 Years Since Ice Age," *The Independent*, August 4, 2005, www.independent.co.uk/news/science/ice-shelf-collapse-was-biggest-for-10000-years-501370.html.

34. Derek V. Ager, *The Nature of the Stratigraphical Record*, 3rd ed. (Chichester, New York, Brisbane, Toronto, Singapore: John Wiley & Sons, 1993), pp. 83–84. Ager, a visiting professor at the University of Toronto when I was a grad student there, is one of my intellectual heroes, and I took to heart one of his articles: "On Seeing the Most Rocks," something any student of geology must do.

35. www.victoria.ac.nz.

36. "Satellites Witness Lowest Arctic Ice Coverage in History," European Space Agency, September 14, 2007, www.esa.int/esaCP/SEMYTC13J6F_index_0.html.

37. Images at www.esa.int/esaEO/SEM6MB9FTLF_index_1.html show the changes in annual Arctic Sea ice.

CHAPTER 6: FLOODING OF COASTAL COUNTRIES AND CITIES

1. Lake IJsselmeer (sometimes translated as Lake IJssel) is a shallow (average depth 15 to 20 feet) but large lake (area of 425 square miles) in the center of Holland. It is the largest lake in Western Europe, and that must really bug the Dutch, who were forced to build it through dam construction. The last thing a country fighting the loss of land wants to do is flood valuable land with a lake. But they had no choice.

Its name comes from the IJssel River, which drains into it via a smaller lake, the Ketelmeer. The lake was created in 1932 when an inland sea, the Zuiderzee, was closed by a 20-mile-wide dam, the Afsluitdijk, as part of a major hydraulic engineering project known as the Zuiderzee Works. The IJsselmeer, a major freshwater reserve, provides water for agriculture and for drinking, and helps control flooding.

2. A severe European windstorm combined with a higher-than-normal tide caused a storm tide that flooded coastal areas. In combination with a tidal surge of the North Sea, the water level locally exceeded 18 feet above mean sea level. The waves overwhelmed sea defenses and caused extensive flooding. Officially, 1,835 people were killed in the Netherlands, mostly in the southwestern province of Zeeland. Others died in England from the same storm.

3. M. G. J. Den Elzen and J. Rotmans, "The Socio-Economic Impact of Sea-Level Rise on the Netherlands: A Study of Possible Scenarios," *Journal of Climate Change* 20, no. 3 (March 1992): 169–195.

4. View a brochure on the project at www.ruimtevoorderivier.nl/files/Files/brochures/EMAB%20PBK%20Engels.pdf.

5. For more information on the dam, visit www.internationalrivers.org/en/china/three-gorges-dam.

6. A good site for maps of areas to be flooded in Bangladesh is http://maps.grida.no/go/graphic/potential-impact-of-sea-level-rise-on-bangladesh.

7. This *don't-miss* article has nothing to do with Kevin Costner's truly awful 1995 epic, which should have been named *Waterloo* for his so-called acting career, but puts much of this chapter into context: Robert D. Kaplan, "Waterworld," *Atlantic*, January–February 2008, www.theatlantic.com/doc/200801/kaplan-bangladesh.

8. Not convinced Bangladesh's situation puts the whole world in danger? Check out the CIA's take on it at https://www.cia.gov/library/publications/the-world-factbook/geos/bg.html.

9. Kaplan, "Waterworld."

10. For more information, see the Organisation for Economic Co-operation and Development report "Development and Climate Change in Bangladesh: Focus on Coastal Flooding and the Sundarbans" at www.oecd.org/dataoecd/46/55/21055658.pdf.

11. See J. P. Doody, ed., *Sand Dune Inventory of Europe*, 2nd ed., National Coastal Consultants and EUCC—the Coastal Union, in association with the IGU Coastal Commission, 2008, www.coastalwiki.org/coastalwiki/Sand_Dunes_in_Europe.

12. See IPCC-4, but also note that others think the figure is much higher. See Michael McCarthy, "Sea Levels Rising Twice as Fast as Predicted," *The Independent*, March 11, 2009, www.independent.co.uk/environment/climate-change/sea-levels-rising-twice-as-fast-as-predicted-1642087.html.

13. See Chapter 5 for estimates of sea level rise over the past thousands of years, or http://climateprogress.org/2009/10/18/science-co2-levels-havent-been-this-high-for-15-million-years-when-it-was-5°-to-10°-warmer-and-seas-were-75-to-120-feet-higher-we-have-shown-that-this-dramatic-rise-in-sea-level-i.

14. This estimate comes from the UN group tracking sea level change and its effects on various nations: "Sea Level Rise and the Vulnerability of Various Coastal Peoples," 2009, www.ehs.unu.edu/file.php?id=652.

15. Summarized in A. Lejour, "Quantifying Four Scenarios for Europe," CPB Netherlands Bureau for Economic Policy Analysis, Den Haag, 2003.

16. Vanessa McKinney, "Sea Level Rise and the Future of the Netherlands," *ICE Case Studies* 12 (May 2007), www1.american.edu/ted/ice/dutch-sea.htm.

17. Ibid.

18. A very prescient film. I had not seen it in years, and the view of a flooded New York is scary. Too bad Spielberg did not end the film with the young robot trapped in the car. See a review by Gary Johnson at www.imagesjournal.com/issue10/reviews/ai.

19. These data came from a bicycling site, www.bikeforums.net/archive/index.php/t-303521.html, and for good reason. Who wants to ride up hills?

20. See the European response: "EU Withdraws Legal Case Against Venice Flood Barriers," Environment News Service, April 15, 2009, www.ens-newswire.com/ens/apr2009/2009-04-15-02.asp. Better yet, check out the great, free interactive site by National Geographic at http://ngm.nationalgeographic.com/2009/08/venice/venice-animation.

21. Ibid.

22. "Once, Venice was protected by this aquatic environment. But now water presents its most serious threat. This month sees the 40th anniversary of a great flood when the lagoon, swelling almost two metres above average sea level, swilled out the walkways and squares. And we just have to face it: Venice is sinking. It has been doing so—at a rate of about 10cm a century—ever since it was built. And things are most certainly not getting any better. Sea levels rise ever more rapidly with global warming. Foundations subside." Rachel Campbell-Johnson, "If You Love Venice, Let Her Die," *Times* (London), June 5, 2006, www.timesonline.co.uk/tol/comment/columnists/rachel_campbell_johnston/article671670.ece.

23. Another great National Geographic site discussing the hoped-for seawalls is http://news.nationalgeographic.com/news/2008/12/photogalleries/Venice-flood-photos/photo3.html.

24. Dominic Standish, "Why We Should Save Venice," *Spiked*, June 15, 2006, www.spiked-online.com/index.php/site/printable/395.

25. http://news.nationalgeographic.com/news/2008/12/photogalleries/Venice-flood-photos/photo3.html.

26. There are lots of sad stories about Michael "Brownie" Brown, director of the Federal Emergency Management Agency, doing a great job and Bush looking down from on high: www.perrspectives.com/blog/archives/000250.htm. For a view from Homeland Security of the future fate of New Orleans, visit www.globalsecurity.org/security/ops/hurricane-risk-new-orleans.htm.

27. The U.S. Geological Survey did a nice job here: E. Robert Thieler, Jeff Williams, and Erika Hammar-Close, "National Assessment of Coastal Vulnerability to Sea-Level Rise," http://woodshole.er.usgs.gov/project-pages/cvi.

28. Median household income: San Francisco listed as $55,221, but nearby Marin county is $71,306. The Shreveport, Louisiana, per capita income in 2000 was just over $17,000, according to the U.S. Census Bureau.

29. An official state plan including this history is at www.aiacc.org/site/docs/bcdc.pdf.

30. San Francisco Bay Conservation and Development Commission, "A Sea Level Rise Strategy for the San Francisco Bay Region," September 2008, www.bcdc.ca.gov/planning/climate_change/SLR_strategy.pdf.

31. "Denver International Airport Construction and Operating Costs," www.colorado.edu/libraries/govpubs/dia.htm.

32. A good site for the effects on ports, which are more endangered than any other human enterprise, is http://ideas.repec.org/p/oec/envaaa/3-en.html. Also see William D. Nordhaus, "Alternative Policies and Sea-Level Rise in the Rice-2009 Model," Cowles Foundation Discussion Paper No. 1716, August 2009, http://cowles.econ.yale.edu/P/cd/d17a/d1716.pdf.

33. The World Resources Institute does a nice job on this: Chris Ward, "Vulnerability and Adaptation to Climate Change in Developing Countries," July 2007, http://earthtrends.wri.org/updates/node/225.

34. The U.S. General Accounting Office (GAO) has this take on the near future: www.gao.gov/new.items/d07863.pdf.

35. The official Canadian government view: www.parl.gc.ca/information/library/PRBpubs/prb0739-e.htm.

36. www.publications.parliament.uk/pa/ld200506/ldselect/ldeconaf/12/12i.pdf.

CHAPTER 7: EXTINCTION?

1. This description comes from http://diversionoz.com/en/gbr-ospreyholmes.htm, a site about diving in Australia: "Osprey Reef: This reef is certainly one of the best in the Coral Sea, it is our favourite reef in the area. Osprey is the most northerly Coral Sea reef, being 350 km away from Cairns. There is a deep-water lagoon in the middle of Osprey Reef, which is great for safe anchorage. There are vertical walls that rise up from more than 1000 metres (3300 feet) to just below the surface! The pelagic action at Osprey is mind blowing. It is a meeting point for the 'big guys.'" I have had the good fortune to take three trips to Osprey Reef, staying about a week each time. Sadly, while I was writing this book, the Undersea Explorer business, which led tours to the reef, folded.

2. I have been involved for three years on an intensive geological investigation of this most magnificent of fossil reefs—of any place or any time. Our work shows quite clearly that a greenhouse extinction caused much loss of species and transition to a microbial reef. See also www.geologynet.com/canning.htm.

3. Much of the work showing the ancient geography of the Canning Basin came from the lifelong work of Dr. Phil Playford.

4. Peter McA. Rees, Mark T. Gibbs, Alfred M. Ziegler, John E. Kutzbach, and Pat J. Behling, "Permian Climates: Evaluating Model Predictions Using Global Paleobotanical Data," *Geology* 27, no. 10 (October 1999): 891–894.

5. Peter Ward, *The Medea Hypothesis* (Princeton, NJ: Princeton University Press, 2009).

6. This very pro-impact point of view reached its height in David M. Raup, *Extinction: Bad Genes or Bad Luck?* (New York: W. W. Norton, 1992). I also described the controversy in *Under a Green Sky*.

7. Peter Ward, "The Greenhouse Extinctions," *Discover*, 1996.

8. Peter Ward, *Under a Green Sky: The Greenhouse Mass Extinctions* (New York: HarperCollins, 2007).

9. Harry L. Bryden et al., "Slowing of the Atlantic Meridional Overturning Circulation at 25°N," *Nature* 438 (December 1, 2005): 655–657.

10. D. Bi, W. F. Budd, A. C. Hirst, and X. Wu, "Collapse and Reorganization of the Southern Ocean Overturning under Global Warming in a Coupled Model," *Geophysical Research Letters* 28, no. 20 (2001): 3927.

11. Bryden et al., "Slowing of the Atlantic Meridional Overturning Circulation at 25°N."

12. Richard B. Alley, *The Two-Mile Time Machine: Ice Cores, Abrupt Climate Change, and Our Future* (Princeton, NJ: Princeton University Press, 2002).

CHAPTER 8: STOPPING CATASTROPHIC SEA LEVEL RISE

1. Tim Flannery, *The Future Eaters: An Ecological History of the Australasian Lands and People* (New York: Grove Press, 2002).

2. "The 160-page 'Copenhagen Climate Treaty,' which has been distributed to negotiators from 192 states, took some of the world's most experienced climate NGOs almost a year to write and contains a full legal text covering all the main elements needed to provide the world with a fair and ambitious agreement that keeps climate change impacts below the unacceptable risk levels identified by most scientists." From the Web site of the World Wildlife Foundation: www.panda.org/about_our_earth/all_publications/?uNewsID =166281.

3. The guest column at RealClimate.org was written by Malte Meinshausen, Reto Knutti, and Dave Frame: www.realclimate.org/index.php/archives/2006/01/can-2c-warming-be-avoided.

4. A good summary of how soil regeneration is used in Pacific island nations with thin tropical soils can be found at www.farmingsolutions.org/successstories/stories.asp?id=141.

5. New York: William Morrow, 2009.

6. New York: Harper Perennial, 2003.

7. A discussion of the Levitt and Dubner chapter on geoengineering is critically examined in a column at RealClimate.org: www.realclimate.org/index.php/archives/2009/10/why-levitt-and-dubner-like-geo-engineering-and-why-they-are-wrong/comment-page-2.

8. The extensive engineering section of this book is rooted in the Royal Society's study "Geoengineering the Climate: Science, Governance and Uncertainty" (http://2020 science.org/2009/09/01/geoengineering-the-climate-a-clear-perspective-from-the-royal-society/#ixzz0VHEnihfi). Specific aspects of the report used here came from the press report of the issuance.

9. The Teller space mesh concept is discussed at http://earth2tech.com/2008/09/01/controversial-globe-changing-measures-could-be-the-only-answer-to-global-warming.

10. This and many other of the technological "fixes" through geoengineering can be found at the very useful Web site Climate Ark: Climate Change and Global Warming Portal, www.climateark.org/geoengineer.

REFERENCES

GENERAL

Barth, M. C., and J. G. Titus, eds. 1984. *Greenhouse Effect and Sea Level Rise: A Challenge for This Generation*. New York: Van Nostrand Reinhold.

Chafee, Senator John H. 1986. "Our Global Environment: The Next Challenge." In *Effects of Changes in Stratospheric Ozone and Global Climate*, ed. J. G. Titus. Washington, DC: U.S. Environmental Protection Agency and United Nations Environment Programme.

Craig, S. G., and K. J. Holmén. 1995. "Uncertainties in Future CO_2 Projections." *Global Biogeochemical Cycles* 9, no. 1:139–152.

Dean, R. G., et al. 1987. *Responding to Changes in Sea Level*. Washington, DC: National Academy Press.

Drewry, D. J., and E. M. Morris. 1992. "The Response of Large Ice Sheets to Climatic Change." *Philosophical Transactions of the Royal Society B* 338: 235–242.

Energy Modeling Forum. 1994. "EMF14: Integrated Assessment of Climate Change Models for which Second Round Scenario Results Have Been Received as of May 10, 1995." Stanford, CA: Energy Modeling Forum. (Available from John P. Weyant, Stanford University, Stanford, CA, 94305-4022.)

Global Climate Change and the Rising Challenge of the Sea. Proceedings of the International Workshop Held on Margarita Island, Venezuela, March 9–13, 1992. Silver Spring, MD: National Ocean Service, U.S. Department of Commerce.

Hoffman, J. S., D. Keyes, and J. G. Titus. 1983. *Projecting Future Sea Level Rise*. Washington, DC: U.S. Environmental Protection Agency.

Intergovernmental Panel on Climate Change. 1994. *Climate Change 1994. Radiative Forcing of Climate Change and an Evaluation of the IPCC IS92 Emission Scenarios*. Cambridge and New York: Cambridge University Press.

_____. 1992. *Climate Change 1992. The Supplementary Report to the IPCC Scientific Assessment*. Cambridge and New York: Cambridge University Press.

_____. 1990. *Climate Change: The IPCC Scientific Assessment*. Cambridge and New York: Cambridge University Press.

Karl, T. R., R. W. Knight, G. Kukla, and J. Gavin. 1995. "Evidence for Radiative Effects of Anthropogenic Sulfate Aerosols in the Observed Climate Record." In *Aerosol Forcing of Climate*, ed. R. Charlson and J. Heintzenberg. Chichester: John Wiley and Sons.

Kerr, R. A. 1995. "Study Unveils Climate Cooling Caused by Pollutant Haze." *Science* 268, no. 5212 (May 12): 802.

Maine, State of. 1995. *Anticipatory Planning for Sea-Level Rise Along the Coast of Maine*. First printing. Augusta: Maine State Planning Office.

_____. 1995. *Anticipatory Planning for Sea-Level Rise Along the Coast of Maine*. Second printing. Washington, DC: U.S. Environmental Protection Agency (Climate Change Division).

Mitchell, J. F. B., R. A. Davis, W. J. Ingram, and C. A. Senior. 1995. "On Surface Temperature, Greenhouse Gases, and Aerosols: Models and Observations." *Journal of Climate* 8, no. 10 (October 1995): 2354–2386.

National Academy of Sciences. 1985. Mark Meier, chairman. *Glaciers, Ice Sheets, and Sea Level*. Washington, DC: National Academy Press.

_____. 1983. *Changing Climate*. Washington, DC: National Academy Press.

_____. 1979. *CO2 and Climate: A Scientific Assessment*. Washington, DC: National Academy Press.

National Research Council. 1987. *Responding to Changes in Sea Level*. R.G. Dean, chairman. Washington, DC: National Academy Press.

Nerem, R. S. 1995. "Global Mean Sea Level Variations from TOPEX/POSEIDON Altimeter Data." *Science* 268, no. 5211 (May 5): 708–710.

Titus, J. G., R. A. Park, S. Leatherman, R. Weggel, M. S. Greene, M. Treehan, S. Brown, C. Gaunt, and G. Yohe. 1991. "Greenhouse Effect and Sea Level Rise: The Cost of Holding Back the Sea." *Coastal Management* 19, no. 3: 171–204.

OTHER CLIMATE-RELATED PAPERS OF INTEREST

Broecker, W. S. 2003. "Does the Trigger for Abrupt Climate Change Reside in the Ocean or in the Atmosphere?" *Science* 300, no. 5625 (June 6): 1519–1522.

Clark, P. U., S. J. Marshall, G. K. C. Clarke, S. W. Hostetler, J. M. Licciardi, and J. T. Teller. 2001. "Freshwater Forcing of Abrupt Climate Change during the Last Glaciation." *Science* 293, no. 5528 (July 13): 283–287.

Clement A. C., M. A. Cane, and R. Seager. 2001. "An Orbitally Driven Tropical Source for Abrupt Climate Change." *Journal of Climate* 14: 2369–2375.

Palmen, E., and C. W. Newton. 1969. *Atmospheric Circulation Systems: Their Structure and Physical Interpretation*. New York and London: Academic Press.

Winton, M. 2003. "On the Climatic Impact of Ocean Circulation." *Journal of Climate* 16: 2875–2889.

MASS EXTINCTION PAPERS OF INTEREST

Alvarez, L. W., W. Alvarez, F. Asaro, and H. V. Michel. 1980. "Extraterrestrial Cause for the Cretaceous-Tertiary Extinction—Experimental Results and Theoretical Interpretation." *Science* 208, no. 4448 (June 6): 1095–1108.

Bambach, R. K., A. H. Knoll, and S. C. Wang. 2004. "Origination, Extinction, and Mass Depletions of Marine Diversity." *Paleobiology* 30, no. 4 (December): 522–542.

Benton, M. J., and R. J. Twitchett. 2003. "How to Kill (Almost) All Life: The End-Permian Extinction Event." *Trends in Ecology and Evolution* 18: 358–365.

Berner, R. A. 2002. "Examination of Hypotheses for the Permo-Triassic Boundary Extinction by Carbon Cycle Modeling." *Proceedings of the National Academy of Sciences* 99: 4172–4177.

Berner, R. A., and Z. Kothavala. 2001. "Geocarb III: A Revised Model of Atmospheric CO_2 Over Phanerozoic Time." *American Journal of Science* 301: 182–204.

Brookfield, M. E., R. J. Twitchett, and C. Goodings. 2003. "Palaeoenvironments of the Permian-Triassic Transition Sections in Kashmir, India." *Palaeogeography, Palaeoclimatology, Palaeoecology* 198: 353–371.

Cerling, T. E. 1992. "Use of Carbon Isotopes in Paleosols as an Indicator of the $P(CO_2)$ of the Paleoatmosphere." *Global Biogeochemical Cycles*: 307–314.

Erwin. D. 2002. "End-Permian Mass Extinctions: A Review." In *Catastrophic Events and Mass Extinctions: Impacts and Beyond*, ed. C. Koeberl and K. G. MacLeod. Boulder, CO: Geological Society of America Special Paper 356.

———. 1994. "The Permo-Triassic Extinction." *Nature* 367: 231–236.

———. 1993. *The Great Paleozoic Crisis: Life and Death in the Permian*. New York: Columbia University Press.

Grice, K., C. Q. Cao, G. D. Love, M. E. Bottcher, R. J. Twitchett, E. Grosjean, R. E. Summons, S. C. Turgeon, W. Dunning, and Y. G. Jin. 2005. "Photic Zone Euxinia During the Permian-Triassic Superanoxic Event." *Science* 307, no. 5710 (February 4): 706–709.

Hsu, K., and J. Mckenzie. 1990. "Carbon Isotope Anomalies at Era Boundaries: Global Catastrophes and Their Ultimate Cause" in V. L. Sharpton and P. D. Ward (eds.), "Global Catastrophes in Earth History: An Interdisciplinary Conference on Impacts, Volcanism, and Mass Mortality." *Geological Society of America Special Paper* 247, 61–70.

Huey, R. B., and P. D. Ward. 2005. "Hypoxia, Global Warming, and Terrestrial Late Permian Extinctions." *Science* 308, no. 5720 (April 15): 398–401.

Isozaki, Y. 1997. "Permo-Triassic Boundary Superanoxia and Stratified Superocean: Records from Lost Deep Sea." *Science* 276, no. 5310 (April 11): 235–238.

Jin, Y.G., Y. Wang, W. Wang, Q. H. Shang, C. Q. Cao, and D. H. Erwin. 2000. "Pattern of Marine Mass Extinction near the Permian-Triassic Boundary in South China." *Science* 289, no. 5478 (July 21): 432–436.

Kidder, D. L., and T. R. Worsley. 2003. "Late Permian Warming, the Rapid Latest Permian Transgression, and the Permo-Triassic Extinction." GSA Annual Meeting, Seattle, WA.

Kiehl J. T., and P. R. Gent. 2004. "The Community Climate System Model, Version Two." *Journal of Climate* 17, no. 19: 3666–3682.

Knoll, A., R. Bambach, D. Canfield, and J. Grotzinger. 1996. "Comparative Earth History and Late Permian Mass Extinction." *Science* 273, no. 5274 (July 26): 452–457.

Kump, L. R., A. Pavlov, and M. A. Arthur. 2005. "Massive Release of Hydrogen Sulfide to the Surface Ocean and Atmosphere during Intervals of Oceanic Anoxia." *Geology* 33, no. 5: 397–400.

Looy, C., W. A. Brugman, D. L. Dilcher, and H. Visscher. 1999. "The Delayed Resurgence of Equatorial Forests after the Permian–Triassic Ecologic Crisis." *Proceedings of the National Academy of Sciences* 96: 13857–13862.

Looy, C., R. J. Twitchett, D. L. Dilcher, J. H. A. Van Konijnenburg-Van Cittert, and H. Visscher. 2001. "Life in the End-Permian Dead Zone." *Proceedings of the National Academy of Sciences* 98: 7879–7883.

Lucas, S. G. 1998. "Global Triassic Tetrapod Biostratigraphy and Biochronology." *Palaeogeography, Palaeoclimatology, Palaeoecology* 143: 347–384.

MacLeod, K. G., R. M. H. Smith, P. L. Koch, and P. D. Ward. 2000. "Timing of Mammal-like Reptile Extinctions across the Permian-Triassic Boundary in South Africa." *Geology* 28: 227–230.

Marshall, C. R. 1995. "Distinguishing Between Sudden and Gradual Extinctions in the Fossil Record: Predicting the Position of the Cretaceous-Tertiary Iridium Anomaly Using the Ammonite Fossil Record on Seymour Island, Antarctica." *Geology* 23: 731–734.

Olsen, P. E., D. V. Kent, H. D. Sues, C. Koeberl, H. Huber, A. Montanari, E. C. Rainforth, S. J. Powell, M. J. Szajna, and B. W. Hartline. 2002. "Ascent of Dinosaurs Linked to an Iridium Anomaly at the Triassic–Jurassic Boundary." *Science* 296 (May 17): 1305–1307.

Palfy, J., A. Demeny, J. Haas, M. Hetenyi, M. J. Orchard, and I. Veto. 2001. "Carbon Isotope Anomaly and Other Geochemical Changes at the Triassic-Jurassic Boundary from a Marine Section in Hungary." *Geology* 29, no. 11: 1047–1050.

Palfy, J., J. K. Mortensen, E. S. Carter, P. L. Smith, R. M. Friedman, and H. W. Tipper. 2000. "Timing the End-Triassic Mass Extinction: First on Land, Then in the Sea?" *Geology* 28, no. 1: 39–42.

Payne, J. L., D. J. Lehrmann, J. Wei, M. J. Orchard, D. P. Schrag, and A. H. Knoll. 2004. "Large Perturbations of the Carbon Cycle During Recovery from the End-Permian Extinction." *Science* 305 (July 23): 506–509.

Raup, D. 1979. "Size of the Permo-Triassic Bottleneck and Its Evolutionary Implications." *Science* 206, no. 4415 (October 12): 217–218.

Retallack, G. J., R. M. H. Smith, and P. D. Ward. 2003. "Vertebrate Extinction across Permian-Triassic Boundary in Karoo Basin, South Africa." *GSA Bulletin* 115: 1133–1152.

Stanley, S. M., and X. Yang. 1994. "A Double Mass Extinction at the End of the Paleozoic Era." *Science* 266, no. 5189 (November 25): 1340–1344.

Ward, P. D., J. Botha, R. Buick, M. O. De Kock, D. H. Erwin, G. Garrison, J. Kirschvink, and R. M. H. Smith. 2005. "Abrupt and Gradual Extinction Among Late Permian Land Vertebrates in the Karoo Basin, South Africa." *Science* 307, no. 5710 (February 4): 709–714.

Ward, P. D., G. H. Garrison, J. W. Haggart, D. A. Kring, and M. J. Beattie. 2004. "Isotopic Evidence Bearing on Late Triassic Extinction Events, Queen Charlotte Islands, British Columbia, and Implications for the Duration and Cause of the

Triassic/Jurassic Mass Extinction." *Earth and Planetary Science Letters* 224, no. 3–4: 589–600.

Ward, P. D., D. R. Montgomery, and R. Smith. 2000. "Altered River Morphology in South Africa Related to the Permian-Triassic Extinction." *Science* 289, no. 5485 (September 8): 1740–1743.

White, R. V. 2002. "Earth's Biggest 'Whodunnit': Unravelling the Clues in the Case of the End-Permian Mass Extinction." *Philosophical Transactions of the Royal Society of London B* 360: 2963–2985.

HOLOCENE ICE SHELF HISTORY

Alley, R. B., P. U. Clark, P. Huybrechts, and I. Joughin. 2005. "Ice-Sheet and Sea-Level Changes." *Science* 310, no. 5747 (October 21): 456–460.

Barber, D. C., et al. 1999. "Forcing of the Cold Event of 8,200 Years Ago by Catastrophic Drainage of Laurentide Lakes." *Nature* 400 (July 22): 344–348.

Bard, E. B., et al. 1996. "Deglacial Sea-Level Record from Tahiti Corals and the Timing of Global Meltwater Discharge." *Nature* 382: 241–244.

Berger, A., and M. F. Loutre. 1991. "Insolation Values for the Climate of the Last 10 Million Years." *Quaternary Science Reviews* 10: 297–317.

Came, R. E., D. W. Oppo, and J. F. McManus. 2007. "Amplitude and Timing of Temperature and Salinity Variability in the Subpolar North Atlantic over the Past 10 k.y." *Geology* 35, no. 4 (April 4): 315–318.

Carlson, A. E., P. U. Clark, G. M. Raisbeck, and E. J. Brook. 2007. "Rapid Holocene Deglaciation of the Labrador Sector of the Laurentide Ice Sheet." *Journal of Climate* 20, no. 20: 5126–5133.

Clark, P. U., E. J. Brook, G. M. Raisbeck, F. Yiou, and J. Clark. 2003. "Cosmogenic 10Be Ages of the Saglek Moraines, Torngat Mountains, Labrador." *Geology* 31, no. 7 (July): 617–620.

Cronin, T. M., et al. 2007. "Rapid Sea Level Rise and Ice Sheet Response to 8,200-Year Climate Event." *Geophysical Research Letters* 34 (October 24).

Dyke, A. S. 2004. In *Quaternary Glaciations—Extent and Chronology*, part 2, vol. 2b, ed. J. Ehlers and P. L. Gibbard, pp. 373–424. Amsterdam: Elsevier.

Fairbanks, R. G. 1989. "A 17,000 Year Glacio-Eustatic Sea Level Record: Influence of Glacial Melting Rates on the Younger Dryas Event and Deep Ocean Circulation." *Nature* 342 (December 7): 637–642.

Gregory, J. M., P. Huybrechts, and S. C. B. Raper. 2004. "Threatened Loss of the Greenland Ice-Sheet." *Nature* 428, no. 616 (April 8).

Hansen, J., et al. 2007. "Dangerous Human-Made Interferences with Climate: A GISS ModelE study. *Atmospheric Chemistry and Physics* 7: 2287–2312.

Hillaire-Marcel, C., and G. Bilodeau. 2000. "Instabilities in the Labrador Sea Water Mass Structure during the Last Climatic Cycle." *Canadian Journal of Earth Sciences* 37, no. 5: 795–809.

Hillaire-Marcel, C., A. deVernal, G. Bilodeau, and A. J. Weaver. 2001. "Absence of Deep-Water Formation in the Labrador Sea During the Last Interglacial Period." *Nature* 410 (April 26): 1073–1077.

Hillaire-Marcel, C., A. deVernal, and D. J. W. Piper. 2007. "Lake Agassiz Final Drainage Event in the Northwest North Atlantic." *Geophysical Research Letters* 34 (August 2).

Intergovernmental Panel on Climate Change. 2007. *Climate Change 2007: The Physical Basis*, Fourth Assessment Report, ed. S. D. Solomon and Qin. M. Manning. Cambridge and New York: Cambridge University Press.

Kaplan, M. R., and A. P. Wolfe. 2006. "Spatial and Temporal Variability of Holocene Temperature in the North Atlantic." *Quaternary Research* 65, no. 2 (March): 223–231.

Kaufman D. S., et al. 2004. "Holocene Thermal Maximum in the Western Arctic (0–180°W)." *Quaternary Science Reviews* 23, no. 5–6 (March): 529–560.

Keigwin, L. D., J. P. Sachs, Y. Rosenthal, and E. A. Boyle. 2005. "The 8200 Year B.P. Event in the Slope Water System, Western Subpolar North Atlantic." *Paleoceanography* 20 (April 15).

Licciardi, J. M., P. U. Clark, J. W. Jenson, and D. R. MacAyeal. 1998. "Deglaciation of a Soft-Bedded Laurentide Ice Sheet." *Quaternary Science Reviews* 17, no. 4: 427–448.

Miller, G. H., A. P. Wolfe, J. P. Briner, P. E. Sauer, and A. Nesje. 2005. "Holocene Glaciation and Climate Evolution of Baffin Island, Arctic Canada." *Quaternary Science Reviews* 24, no. 14–15 (August): 1703–1721.

Mitchell, J. F. B., N. S. Grahame, and K. J. Needham. 1988. "Climate Simulations for 9000 Years Before Present: Seasonal Variations and Effect of the Laurentide Ice Sheet." *Journal of Geophysical Research Atmospheres* 93: 8283–8303.

Paterson, W. S. B. 1994. *The Physics of Glaciers*. Oxford, UK: Butterworth-Heinemann.

Peltier, W. R. 2004. "Global Glacial Isostasy and the Surface of the Ice-Age Earth: The ICE-5G (VM2) Model and GRACE." *Annual Review of Earth and Planetary Sciences* 32: 111–149 (2004).

Pollard, D., J. C. Bergengren, L. M. Stillwell-Soller, B. Felzer, and S. L. Thompson. 1998. "Climate Simulations for 10,000 and 6,000 Years BP." *Paleoclimates* 2: 183–218.

Ridley, J. K., P. Huybrechts, J. M. Gregory, and J. A. Lowe. 2005. "Elimination of the Greenland Ice Sheet in a High CO_2 Climate." *Journal of Climate* 18, no. 17: 3409–3427.

Schmidt, G. A., A. N. LeGrande, and G. Hoffmann. 2007. "Water Isotope Expressions of Intrinsic and Forced Variability in a Coupled Ocean-Atmosphere Model." *Journal of Geophysical Research Atmospheres* 112 (May 17).

Wild, M., P. Calanca, S. C. Scherrer, and A. Ohmura. 2003. "Effects of Polar Ice Sheets on Global Sea Level in High-Resolution Greenhouse Scenarios." *Journal of Geophysical Research* 108 (March 13).

Yu, S.-Y., B. E. Berglund, P. Sandgren, and K. Lambeck. 2007. "Evidence for a Rapid Sea-Level Rise 7600 Yr Ago." *Geology* 35, no. 10 (October): 891–894.

SEA-LEVEL PROJECTIONS

Associated Press. 1985. "Doubled Erosion Seen for Ocean City." *Washington Post*, November 14 (Maryland Section).

Barth, M. C., and J. G. Titus, eds. 1984. *Greenhouse Effect and Sea Level Rise: A Challenge for This Generation*. New York: Van Nostrand Reinhold.

Chafee, Senator John H. 1986. "Our Global Environment: The Next Challenge." In *Effects of Changes in Stratospheric Ozone and Global Climate*, ed. J. G. Titus. Washington, DC: U.S. Environmental Protection Agency and United Nations Environment Programme.

Craig, S. G., and K. J. Holmén. 1995. "Uncertainties in Future CO_2 Projections." *Global Biogeochemical Cycles* 9: 139–152.

Dean, R. G., et al. 1987. *Responding to Changes in Sea Level*. Washington, DC: National Academy Press.

Drewry, D. J., and E. M. Morris. 1992. "The Response of Large Ice Sheets to Climatic Change." *Philosophical Transactions of the Royal Society of London B* 338: 235–242.

Energy Modeling Forum. 1994. "EMF14: Integrated Assessment of Climate Change Models for Which Second Round Scenario Results Have Been Received as of May 10, 1995." Stanford, CA: Energy Modeling Forum. (Available from John P. Weyant, Stanford University, Stanford, CA 94305-4022.)

Hoffman, J. S., D. Keyes, and J. G. Titus. 1983. *Projecting Future Sea Level Rise*. Washington, DC: U.S. Environmental Protection Agency.

Hoffman, J. S., et al. 1982 (draft). *Projecting Future Sea Level Rise*. Washington, DC: U.S. Environmental Protection Agency.

Intergovernmental Panel on Climate Change. 1995. *Global Climate Change and the Rising Challenge of the Sea*. Proceedings of the international workshop held on Margarita Island, Venezuela, March 9–13, 1992. Silver Spring, MD: National Ocean Service, U.S. Department of Commerce.

————. 1994. *Climate Change 1994. Radiative Forcing of Climate Change and an Evaluation of the IPCC IS92 Emission Scenarios*. Cambridge and New York: Cambridge University Press.

————. 1992. *Climate Change 1992. The Supplementary Report to the IPCC Scientific Assessment*. Cambridge and New York: Cambridge University Press.

————. 1990. *Climate Change: The IPCC Scientific Assessment*. Cambridge and New York: Cambridge University Press.

Karl, T. R., R. W. Knight, G. Kukla, and J. Gavin. 1995. "Evidence for Radiative Effects of Anthropogenic Sulfate Aerosols in the Observed Climate Record." In *Aerosol Forcing of Climate*, ed. R. Charlson and J. Heintzenberg. Chichester: John Wiley and Sons.

Kerr, R. A. 1995. "Study Unveils Climate Cooling Caused by Pollutant Haze." *Science* 268, no. 5212 (May 12): 802.

Maine, State of. 1995. *Anticipatory Planning for Sea-Level Rise Along the Coast of Maine*. First printing. Augusta: Maine State Planning Office.

————. 1995. *Anticipatory Planning for Sea-Level Rise Along the Coast of Maine*. Second printing. Washington, DC: U.S. Environmental Protection Agency, Climate Change Division.

Mitchell, J. F. B., R. A. Davis, W. J. Ingram, and C. A. Senior. 1995. "On Surface Temperature, Greenhouse Gases, and Aerosols: Models and Observations." *Journal of Climate* 8, no. 10 (October): 2364–2386.

National Research Council. 1987. *Responding to Changes in Sea Level*. R. G. Dean, chairman. Washington, DC: National Academy Press.

THERMOHALINE CIRCULATION CHANGES

Bi, D., W. F. Budd, A. C. Hirst, and X. Wu. 2001. "Collapse and Reorganization of the Southern Ocean Overturning under Global Warming in a Coupled Model." *Geophysical Research Letters* 28, no. 20: 3927.

Clark, P. U., N. G. Pisias, T. F. Stocker, and A. J. Weaver. 2002. "The Role of the Thermohaline Circulation in Abrupt Climate Change." *Nature* 415 (February 21): 863.

Curry, R., B. Dickson, and I. Yashayaev. 2003. "A Change in the Freshwater Balance of the Atlantic Ocean Over the Past Four Decades." *Nature* 426 (December 18/25): 826.

Dickson, B., I. Yashayaev, J. Meincke, B. Turrell, S. Dye, and J. Holfort. 2002. "Rapid Freshening of the Deep North Atlantic Ocean over the Past Four Decades." *Nature* 416 (April 25): 832–837.

Fichefet, T., and M. A. M. Maqueda. 1997. "Sensitivity of a Global Sea Ice Model to the Treatment of Ice Thermodynamics and Dynamics." *Journal of Geophysical Research* 102: 12609.

Ganachaud, A., and C. Wunsch. 2000. "Improved Estimates of Global Ocean Circulation, Heat Transport and Mixing from Hydrographic Data." *Nature* 408 (November 23): 453–457.

Hakkinen, S., and P. B. Rhines. 2004. "Decline of Subpolar North Atlantic Circulation During the 1990s." *Science* 304, no. 5670 (April 23): 555–559.

Hirschi, J., J. Baehr, J. Marotzke, J. Stark, S. Cunningham, and J. O. Beismann. 2003. "A Monitoring Design for the Atlantic Meridional Overturning Circulation." *Geophysical Research Letters* 30, no. 7: 1413.

Hsieh W. W., and K. Bryan. 1996. "Redistribution of Sea Level Rise Associated with Enhanced Greenhouse Warming: A Simple Model Study." *Climate Dynamics* 12: 535–544.

Intergovernmental Panel on Climate Change. 2001. *Climate Change 2001: The Scientific Basis. Contribution of Working Group I to the Third Assessment Report of the Intergovernmental Panel on Climate Change.* Cambridge: Cambridge University Press.

Johnson, H. L., and D. P. Marshall. 2004. "Global Teleconnections of Meridional Overturning Circulation Anomalies." *Journal of Physical Oceanography* 34: 1702.

————. 2002. A Theory for the Surface Atlantic Response to Thermohaline Variability." *Journal of Physical Oceanography* 32: 1121.

Kalnay, E., M. Kanamitsu, R. Kistler, W. Collins, D. Deaven, L. Gandin, M. Iredell, S. Saha, G. White, J. Woollen, Y. Zhu, M. Chelliah, W. Ebisuzaki, W. Higgins, J. Janowiak, K. C. Mo, C. Ropelewski, J. Wang, A. Leetmaa, R. Reynolds, R. Jenne, and D. Joseph. 1996. "The NCEP/NCAR 40-Year Reanalysis Project." *Bulletin of the American Meteorological Society* 77, no. 3: 437–471.

Knutti, R., and T. F. Stocker. 2000. "Influence of the Thermohaline Circulation on Projected Sea Level Rise." *Journal of Climate* 13: 1997.

Lombard, A., A. Cazenave, K. DoMinh, C. Cabanes, and R. S. Nerem. 2004. "Thermosteric Sea Level Rise for the Past 50 Years: Comparison with Tide Gauges and Inference on Water Mass Contribution." *Global and Planetary Change* 48, no. 4: 303–312.

Manabe, S., and R. J. Stouffer. 1999. "Are Two Modes of Thermohaline Circulation stable?" *Tellus* 51A: 400.

_____. 1993. "Century-Scale Effects of Increased Atmospheric CO_2 on the Ocean-Atmosphere System." *Nature* 364, no. 6434: 215–218.

_____. 1988. "Two Stable Equilibria of a Coupled Ocean-Atmosphere Model." *Journal of Climate* 1: 841.

Montoya, M., A. Griesel, A. Levermann, J. Mignot, M. Hofmann, A. Ganopolski, and S. Rahmstorf. 2004 (under revision). "The Earth System Model of Intermediate Complexity CLIMBER-3a. Part I: Description and Performance for Present Day Conditions."

CLIMATE DYNAMICS

NOAA. 1992–1995. *Altimeter Gridded Sea Level Analysis: Sea Surface Height Anomaly.* NOAA TOPEX/POSEIDON.

Peterson, B. J., R. M. Holmes, J. W. McClelland, J. Vörösmarty, R. B. Lammers, A. I. Shiklomanov, I. A. Shiklomanov, and S. Rahmstorf. 2002. "Increasing River Discharge to the Arctic Ocean." *Science* 298, no. 5601 (December 13): 2171–2173.

Petoukhov, V., A. Ganopolski, V. Brovkin, M. Claussen, A. Eliseev, C. Kubatzki, and S. Rahmstorf. 2000. "CLIMBER-2: A Climate System Model of Intermediate Complexity Part I: Model Description and Performance for Present Climate." *Climate Dynamics* 16, no. 1: 1–17.

Rahmstorf, S. 2002. "Ocean Circulation and Climate During the Past 120,000 Years." *Nature* 419 (September 12): 207–214.

_____. 1999. "Shifting Seas in the Greenhouse?" *Nature* 399 (June 10): 523–524.

Rahmstorf, S., and A. Ganopolski. 1999. "Long-Term Global Warming Scenarios Computed with an Efficient Coupled Climate Model." *Climatic Change* 43: 353.

Schaeffer, M., F. M. Selten, J. D. Opsteegh, and H. Goose. 2002. "Intrinsic Limits to Predictability of Abrupt Regional Climate Change in IPCC SRES Scenarios." *Geophysical Research Letters* 29, no. 16: 1767.

Schwartz, P., and D. Randall. 2003. "An Abrupt Climate Change Scenario and Its Implications for United States National Security." Pentagon Report.

Seidov, D., B. J. Haupt, E. J. Barron, and M. Maslin. 2001. "Ocean Bi-Polar Seesaw and Climate: Southern Versus Northern Meltwater Impacts." In *The Oceans and Rapid Climate Change: Past, Present, and Future*, ed. D. Seidov, B. J. Haupt, and M. Maslin. Geophysical Monograph Series vol. 126: 147.

Stocker, T. F., and A. Schmittner. 1997. "Influence of CO_2 Emission Rates on the Stability of the Thermohaline Circulation." *Nature* 388 (August 28): 862–865.

Stouffer, R. J., and S. Manabe. 2003. "Equilibrium Response of Thermohaline Circulation to Large Changes in Atmospheric CO_2 Concentration." *Climate Dynamics* 20: 759.

Talley, and L. D., J. L. Reid, and P. E. Robbins. 2003. "Data-Based Meridional Overturning Streamfunctions for the Global Ocean." *Journal of Climate* 16: 3213.

Vellinga, M., and R. A. Wood. 2002. "Global Climatic Impacts of a Collapse of the Atlantic Thermohaline Circulation." *Climatic Change* 54: 251.

Wijffels, S. E., R. W. Schmitt, H. L. Bryden, and A. Stigebrandt. 1992. "Transport of Freshwater by the Oceans." *Journal of Physical Oceanography* 22, no. 2: 155.

Wood, R. A., A. B. Keen, J. F. B. Mitchell, and J. M. Gregory. 1999. "Changing Spatial Structure of the Thermohaline Circulation in Response to Atmospheric CO_2 Forcing in a Climate Model." *Nature* 399 (June 10):572.

LANDS ENDANGERED BY SEA LEVEL RISE

Armentano, T. V., R. A. Park, and C. L. Cloonan. 1988. "Impacts on Coastal Wetlands Throughout the United States." In *Greenhouse Effect, Sea Level Rise, and Coastal Wetlands*, ed. J. G. Titus. Washington, DC: U.S. Environmental Protection Agency.

Asthana, V., S. Surendra, and S. Venkatesan. 1992. "Assessment of the Vulnerability of Orissa and West Bengal, India, to Sea Level Rise." In Coastal Zone Management Subgroup, Intergovernmental Panel on Climate Change. *Global Climate Change and the Rising Challenge of the Sea*. Available from the National Technical Information Service.

Awosika, L. A., G. T. French, R. J. Nichols, and C. E. Ibe. 1992. "The Impacts of Sea Level Rise on the Coastline of Nigeria." In Coastal Zone Management Subgroup, Intergovernmental Panel on Climate Change. *Global Climate Change and the Rising Challenge of the Sea*. Available from the National Technical Information Service.

Bates, T. 1999. "Sink or Swim." *Asbury Park Press*, Millennium Section at 1. July 4. Also found online at www.injersey.com/2000/story/1,2297,195727,00.html.

Bruun, P. 1962. "Sea Level Rise as a Cause of Shore Erosion." *Journal of Waterways and Harbors Division* (ASCE) 88: 116–130.

Camber, G. 1992. "Assessment of the Vulnerability of Coastal Areas in Coastal Zone Management Subgroup, Intergovernmental Panel on Climate Change." In *Global Climate Change and the Rising Challenge of the Sea*. Available from the National Technical Information Service.

Dunbar, J. B., L. D. Britsch, and E. B. Kemp. 1992. *Land Loss Rates: Report 3, Louisiana Coastal Plain*. New Orleans: U.S. Army Corps of Engineers.

Eastman, J. R., and S. Gold. 1996. "Spatial Information Systems and Assessment of the Impacts of Sea Level Rise." In E. Lorup and J. Strobol. "IDRISI GIS 96 ' Salzburger Geographische Materialien, Heft 25." Selbstverlag des Instituts für. Geographie der Universität Salzburg. Also found at www.sbg.ac.at/geo/idrisi/idrgis96/eastman1/slrise.htm.

El-Raey, M., S. Nasr, O. Frihy, S. Desouki, and Kh. Dewidar. 1995. "Potential Impacts of Accelerated Sea-Level Rise on Alexandria Governorate, Egypt." *Journal of Coastal Research*, Special Issue 14: 190–204.

Environmental Protection Agency. 1995. *The Probability of Sea Level Rise*. Washington, DC: U.S. Environmental Protection Agency. Also available at www.epa.gov/globalwarming/publications/impacts/sealevel/probability.html.

_____. 1989. *The Potential Impacts of Global Climate Change on the United States*. Washington, DC: U.S. Environmental Protection Agency.

_____. 1983. *Projecting Future Sea Level Rise*. Washington, DC: Strategic Studies Staff, U.S. Environmental Protection Agency.

Federal Emergency Management Agency, Federal Insurance Administration. 1991. *Projected Impact of Relative Sea Level Rise on the National Flood Insurance Program*. Washington, DC: Federal Emergency Management Agency.

Han, M., J. Hou, and L. Wu. 1995. "Potential Impacts of Sea Level Rise on China's Coastal Environment and Cities: A National Assessment." *Journal of Coastal Research*, Special Issue 14: 79–95.

Huq, A, S. I. Ali, and A. A. Rahman. 1995. "Sea Level Rise and Bangladesh: A Preliminary Analysis." *Journal of Coastal Research*, Special Issue 14: 44–53.

Intergovernmental Panel on Climate Change. 1998. *The Regional Effects of Climate Change*. New York and Cambridge: Cambridge University Press.

———. 1996a. *Climate Change 1995: The Science of Climate Change*. New York and Cambridge: Cambridge University Press.

———. 1996b. *Climate Change 1995: Impacts, Adaptations, and Mitigation of Climate Change*. New York and Cambridge: Cambridge University Press.

———. 1990. *The IPCC Scientific Assessment*. Cambridge and New York: Cambridge University Press.

Jensen, R. 1985. "Stopping the Sinking." *Texas Water Resources Institute* 11: 3.

Jogoo, V. 1992. "Assessment of the Vulnerability of Mauritius to Sea Level Rise." In Coastal Zone Management Subgroup, Intergovernmental Panel on Climate Change. *Global Climate Change and the Rising Challenge of the Sea*. Available from the National Technical Information Service.

Kana, T. W., J. Michel, M. O. Hayes, and J. R. Jensen. 1984. "The Physical Impact of Sea Level Rise in the Area of Charleston, South Carolina." In M. C. Barth and J. G. Titus, eds. *Greenhouse Effect and Sea Level Rise: A Challenge for This Generation*. New York: Van Nostrand Reinhold.

Kana, T. W., B. J. Baca, and M. L. Williams. 1988. "Charleston Case Study." In *Greenhouse Effect, Sea Level Rise, and Coastal Wetlands*, ed. J. G. Titus. Washington, DC: U.S. Environmental Protection Agency. Also available online at www.epa.gov/globalwarming/publications/impacts/sealevel/index.html.

———. 1986. "Potential Impacts of Sea Level Rise on Wetlands Around Charleston, South Carolina." Washington, DC: U.S. Environmental Protection Agency.

Kay, R., I. Eliot, and G. Klem. 1992. "Assessment of the Vulnerability of Geographe Bay, Australia, to Sea Level." *Maps of Lands Vulnerable to Sea Level Rise: Modeled Elevations Along the U.S. Atlantic and Gulf Coasts Rise*. In Coastal Zone Management Subgroup, Intergovernmental Panel on Climate Change. *Global Climate Change and the Rising Challenge of the Sea*. Available from the National Technical Information Service.

Kearney, M. S., and J. C. Stevenson. 1985. "Sea Level Rise and Marsh Vertical Accretion Rates in Chesapeake Bay." In *Coastal Zone '85*, ed. O. T. Magoon et al. New York: American Society of Civil Engineers.

Keeling, C. D., R. B. Bacastow, A. F. Carter, S. C. Piper, T. P. Whorf, M. Heinmann, W. G. Mook, and H. Roeloffzen. 1989. "A Three Dimensional Model of Atmospheric Transport Based on Observed Winds: Analysis of Observational Data." In *Aspects of Climate Variability in the Pacific and Western Americas*, ed. D. H. Peterson. Geophysical Monograph 55: 165–236. Washington, DC: AGU.

Keeling, C. D., T. P. Whorf, M. Wahlen, and J. van der Plicht. 1995. "Interannual Extremes in the Rate of Atmospheric Carbon Dioxide since 1980." *Nature* 375: 666–670.

Landsea, C. W. 1999. "FAQ : Hurricanes, Typhoons, and Tropical Cyclones. Part G: Tropical Cyclone Climatology." Miami: National Hurricane Center. Also on the National Hurricane Center Web site, www.aoml.noaa.gov/hrd/tcfaq/tcfaqG.html#G9.

Leake, S. A. 1997. "Land Subsidence From Ground-Water Pumping." Reston, VA: U.S. Geological Survey. Also online at http://geochange.er.usgs.gov/sw/changes/anthropogenic/subside.

Leatherman, S. P. 1984. "Coastal Geomorphic Responses to Sea Level Rise in and Around Galveston, Texas." In *Greenhouse Effect and Sea Level Rise: A Challenge for This Generation*, ed. M. C. Barth and J. G. Titus. New York: Van Nostrand Reinhold.

Leatherman, S. P., R. J. Nichols, and K. C. Dennis. 1995. "Aerial Videotape-Assisted Analysis: A Cost-Effective Approach to Assess Sea Level Rise Impacts." *Journal of Coastal Research*, Special Issue 14: 15–25.

Louisiana Wetland Protection Panel. 1987. *Saving Louisiana's Wetlands: The Need for a Long-Term Plan of Action*. Washington, DC: U.S. Environmental Protection Agency.

Lyle, S. D., L. H. Hickman, and H. A. Debaugh. 1986. "Sea Level Variations in the United States." Rockville, MD: National Oceanographic and Atmospheric Administration.

Mercer, J. H. 1978. "West Antarctic Ice Sheet and CO_2 Greenhouse Effect: A Threat of Disaster?" *Nature* 271: 321–325.

Mimura, N., M. Isobe, and Y. Hosakawa. 1992. "Impacts of Sea Level Rise on Japanese Coastal Zones and Response Strategies." In Intergovernmental Panel on Climate Change. 1996. *Climate Change 1995: The Science of Climate Change*. New York and Cambridge: Cambridge University Press.

Morton, R., and T. Paine. 1990. "Coastal Land Loss in Texas, an Overview." *Gulf Coast Assessment of Geological Societies Transactions* 40: 625–634.

National Academy of Sciences. 1985. *Glaciers, Ice Sheets, and Sea Level*. Polar Research Board, Committee on Glaciology, Mark Meier, chairman. Washington, DC: National Academy Press.

———. 1983. *Changing Climate*. Washington, DC: National Academy Press.

National Environmental Trust. 1998. *The Massachusetts Coast: On the Front Lines of Global Warming*. Washington, DC: National Environmental Trust. Available online at www.envirotrust.com/PDF/MassCoast.pdf (April 1, 1999).

National Geodetic Survey. 1998. *National Height Modernization Study*. Written by Dewberry and Davis, and Psomas and Associates. Silver Spring, MD: National Geodetic Survey.

National Oceanic and Atmospheric Administration. 1999. *NOAA's Medium Resolution Digital Vector Shoreline*. Silver Spring, MD: Office of Ocean Resources Conservation and Assessment.

National Park Service. 1999. "Cape Hatteras Lighthouse Relocation Articles and Images." www.nps.gov/caha/lrp.htm.

Niang, I., K. C. Dennis, and R. J. Nichols. 1992. "The Impacts of Sea Level Rise on the Coastline of Senegal." In Coastal Zone Management Subgroup, Intergovernmental Panel on Climate Change. *Global Climate Change and the Rising Challenge of the Sea*. Available from the National Technical Information Service.

Park, R. A., M. S. Treehan, P. W. Mausel, and R. C. Howe. 1989. "The Effects of Sea Level Rise on U.S. Coastal Wetlands." In *Potential Effects of Global Climate Change on the United States*. Washington, DC: Environmental Protection Agency.

Pilkey, Orrin H., William J. Neal, Stan Riggs, Deborah Pilkey, and Craig A. Webb. 1998. *The North Carolina Shore and Its Barrier Islands: Restless Ribbons of Sand (Living With the Shore)*. Durham, NC: Duke University Press.

Sallenger, A., et al. 1999. "Airborne Laser Study Quantifies El Nino–Induced Coastal Change." *EOS Transactions* 80, no. 8: 89–93.

Schnack, E. J., K. C. Dennis, R. J. Nichols, and F. Mouzo. 1992. "Impacts of Sea Level Rise on the Coast of Argentina." In Coastal Zone Management Subgroup, Intergovernmental Panel on Climate Change. *Global Climate Change and the Rising Challenge of the Sea*. Available from the National Technical Information Service.

Schneider, S. H., and R. S. Chen. 1980. "Carbon Dioxide Flooding: Physical Factors and Climatic Impact." *Annual Review of Energy* 5: 107–140.

Titus, J. G. 1998. "Rising Seas, Coastal Erosion, and the Takings Clause: How to Save Wetlands and Beaches Without Hurting Property Owners." *Maryland Law Review* 57: 1281–1398. Also available online at www.epa.gov/globalwarming/publications/ impacts/sealevel/index.html.

Titus, J. G., and M. Greene. 1989. "An Overview of the Nationwide Impacts of Sea Level Rise." In *Potential Impacts of Global Climate Change on the United States*. Washington, DC: U.S. Environmental Protection Agency.

Titus, J. G. and V. Narayanan. 1996. "The Risk of Sea Level Rise: A Delphic Monte Carlo Analysis in which Twenty Researchers Specify Subjective Probability Distributions for Model Coefficients within their Respective Areas of Expertise." *Climatic Change* 33: 151–212. Also available online at www.epa.gov/globalwarming/publications/ impacts/sealevel/index.html.

Titus, J. G., R. A. Park, S. Leatherman, R. Weggel, M. S. Greene, M. Treehan, S. Brown, C. Gaunt, and G. Yohe. 1991. "Greenhouse Effect and Sea Level Rise: The Cost of Holding Back the Sea." *Coastal Management*. 19, no. 3: 171–204. Also available online at www.epa.gov/globalwarming/publications/impacts/sealevel/index.html.

Volonte, C. R., and R. J. Nichols. 1995. "Sea Level Rise and Uruguay: Potential Impacts and Responses." *Journal of Coastal Research*, Special Issue 14: 285–302.

Yohe, G. 1990. "The Cost of Not Holding Back the Sea." *Coastal Management* 18: 403–432.

Yohe, G., J. Neumann, P. Marshall, and H. Ameden. 1996. "The Economic Cost of Greenhouse-Induced Sea-Level Rise for Developed Property in the United States." *Climatic Change* 32: 387–410.

ANTARCTICA AND ITS CIRCULATION AND GLACIAL HISTORY

Barker, P. F., and J. Burrell. 1977. "The Opening of the Drake Passage." *Marine Geology* 25: 15–34.

Barrett, P. J. 1994. "Progress Towards a Cenozoic Antarctic Glacial History." *Terra Antarctica* 1: 247–248.

Barrows, T. T., S. Jugins, P. DeDeckker, J. Thiede, and J. I. Martinez. 2000. "Sea-Surface Temperatures of the Southwest Pacific Ocean during the Last Glacial Maximum." *Paleoceanography* 15: 95–109.

Berger, W. H., T. Bickert, M. K. Yasuda, and G. Wefer. 1996. "Reconstruction of Atmospheric CO_2 from the Deep-Sea Record of Ontong Java Plateau: The Milankovitch Chron." *Geologische Rundschau* 85, no. 3 (September): 466–495.

Cande, S. C., J. M. Stock, R. D. Muller, and T. Ishihara. 2000. "Cenozoic Motion between East and West Antarctica." *Nature* 404 (March 9): 145–150.

Diester-Haass, L., and R. Zahn. 1996. "Eocene-Oligocene Transition in the Southern Ocean: History of Water Mass Circulation and Biological Productivity." *Geology* 24, no. 2 (February): 163–166.

Garner, D. M. 1959. "The Subtropical Convergence in New Zealand Waters." *New Zealand Journal of Geology and Geophysics* 2: 315–337.

Hall, R. 1996. "Reconstructing Cenozoic SE Asia." In *Tectonic Evolution of Southeast Asia*, ed. R. Hall and D. J. Blundell. Geological Society Special Publication London, 106: 153–184.

Haq, B. U., J. Hardenbol, and P. R. Vail. 1987. "Chronology of Fluctuating Sea Levels since the Triassic." *Science* 235, no. 4793 (March 6): 1156–1167.

Heusser, L. E., and G. Van de Geer. 1994. "Direct Correlation of Terrestrial and Marine Paleoclimate Records from Four Glacial-Interglacial Cycles; DSDP Site 594, Southwest Pacific." In Murray Wallace, C.V., ed., "Quaternary Marine and Terrestrial Records in Australasia; Do They Agree?" *Quaternary Science Reviews* 13: 273–282.

Howard, W. R., and W. L. Prell. 1994. "Late Quaternary Carbonate Production and Preservation in the Southern Ocean: Implications for Oceanic and Atmospheric Carbon Cycling." *Paleoceanography* 9: 453–482.

———. 1992. "Late Quaternary Surface Circulation of the Southern Indian Ocean and Its Relationship to Orbital Variations." *Paleoceanography* 7: 79–117.

Imbrie, J., E. A. Boyle, S. C. Clemens, A. Duffy, W. R. Howard, G. Kukla, J. Kutzbach, D. G. Martinson, A. McIntyre, A. C. Mix, B. Molfino, J. J. Morley, L. C. Peterson, N. G. Pisias, W. L. Prell, M. E. Raymo, N. J. Shackleton, and J. R. Toggweiler. 1993. "On the Structure and Origin of Major Glaciation Cycles, 2. The 100,000-Year Cycle." *Paleoceanography* 8: 698–736.

———.1992. "On the Structure and Origin of Major Glaciation Cycles, 1. Linear Responses to Milankovitch Forcing." *Paleoceanography* 7: 701–738.

Kennett, J. P. 1977. "Cenozoic Evolution of Antarctic Glaciation, the Circum-Antarctic Ocean, and Their Impact on Global Paleoceanography." *Journal of Geophysical Research* 82: 3843–3860.

Kennett, J. P., R. E. Houtz, P. B. Andrews, A. R. Edwards, V. A. Gostin, M. Hajos, M. A. Hampton, D. G. Jenkins, S. V. Margolis, A. T. Ovenshine, and K. Perch-Nielsen. 1974. "Development of the Circum-Antarctic Current." *Science* 186, no. 4159 (October 11): 144–147.

Kennett, J. P., and N. J. Shackleton. 1976. "Oxygen Isotopic Evidence for the Development of the Psychrosphere 38 Myr ago." *Nature* 260: 513–515.

Lawver, L. A., L. M. Gahagan, and M. F. Coffin. 1992. "The Development of Paleoseaways around Antarctica." In J. P. Kennett and D. A. Warnke, eds., "The Antarctic Paleoenvironment: A Perspective on Global Change." *Antarctic Research Series* 56: 7–30.

McKerron, A. J., V. L. Dunn, R. M. Fish, C. R. Mills, and S. K. van der Linden-Dhont. 1998. "Bass Strait's Bream B Reservoir Development: Success through a Multi-Functional Team Approach." *APEA Journal* 38: 13–35.

Miller, K. G., R. G. Fairbanks, and G. S. Mountain. 1987. "Tertiary Oxygen Isotope Synthesis, Sea-Level History, and Continental Margin Erosion." *Paleoceanography* 2: 1–19.

Miller, K. G., J. D. Wright, and R. G. Fairbanks. 1991. "Unlocking the Ice House: Oligocene Miocene Oxygen Isotopes, Eustasy, and Margin Erosion." *Journal of Geophysical Research* 96: 6829–6848.

Mohr, B. A. R. 1990. "Eocene and Oligocene Sporomorphs and Dinoflagellate Cysts," from J. C. Mutter, K. A. Hegarty, S. C. Cande, and J. K. Weissel. 1985. "Breakup between Australia and Antarctica: A Brief Review in the Light of New Data." *Tectonophysics* 114: 255–279.

Orsi, A. H., T. Whitworth III, and W. D. Nowlin Jr. 1995. "On the Meridional Extent and Fronts of the Antarctic Circumpolar Current." *Deep-Sea Research* 42: 641–673.

Pyle, D. G., D. M. Christie, J. J. Mahoney, and R. A. Duncan. 1995. "Geochemistry and Geochronology of Ancient Southeast Indian and Southwest Pacific Seafloor." *Journal of Geophysical Research* 100: 22261–22282.

Shackleton, N. J., and J. P. Kennett. 1975. "Paleotemperature History of the Cenozoic and the Initiation of Antarctic Glaciation: Oxygen and Carbon Isotope Analyses in DSDP Sites 277, 279, and 281." In J. P. Kennett, R. E. Houtz, et al., Initial Reports DSDP, 29: Washington, DC (U.S. Govt. Printing Office), pp. 743–755.

Shipboard Scientific Party, in press. Proc. ODP, Initial Reports, 188: College Station, Texas (Ocean Drilling Program).

Smith, G. C. 1985. "Bass Basin Geology and Petroleum Exploration." In R. C. Glenie, ed., "Second South-Eastern Australia Oil Exploration Symposium: Technical Papers Presented at the PESA Symposium." Petroleum Exploration Society of Australia: 257–284.

Sloan, E. D. Jr. 1998. "Clathrate Hydrates of Natural Gases." 2nd ed. New York: Marcel Dekkenr Inc., p. 705.

Stott, L. D., J. P. Kennett, N. J. Shackleton, and R. M. Corfield. 1990. "The Evolution of Antarctic Surface Waters during the Paleogene: Inferences from the Stable Isotopic Composition of Planktonic Foraminifers, ODP Leg 113." In P. F. Barker, J. P. Kennett, et al., Proc. ODP, Sci. Results, 113: College Station, Texas (Ocean Drilling Program), 849–863.

Trenberth, K. E., and A. Solomon. 1994. "The Global Heat Balance: Heat Transports in the Atmosphere and Ocean." *Climate Dynamics* 10: 107–134.

Wei, W. 1991. "Evidence for an Earliest Oligocene Abrupt Cooling in the Surface Waters of the Southern Ocean." *Geology* 19: 780–783.

Weissel, J. K., and D. E. Hayes. 1972. "Magnetic Anomalies in the Southeast Indian Ocean." In *Antarctic Oceanology (Vol. 2): The Australian-New Zealand Sector*, ed. D. E. Hayes. Antarctic Research Series 19: 165–196.

Wright, J. D., K. G. Miller, and R. G. Fairbanks. 1991. "Evolution of Modern Deepwater Circulation; Evidence from the Late Miocene Southern Ocean." *Paleoceanography* 6: 275–290.

Zachos, J. C., J. R. Breza, and S. W. Wise. 1992. "Early Oligocene Ice-Sheet Expansion on Antarctica: Stable Isotope and Sedimentological Evidence from Kerguelen Plateau, Southern Indian Ocean." *Geology* 20: 569–573.

Zachos, J. C., K. C. Lohmann, J. C. G. Walker, and S. W. Wise Jr. 1993. "Abrupt Climate Change and Transient Climates during the Paleogene: A Marine Perspective." *Journal of Geology* 101: 191–213.

Zachos, J. C., L. D. Stott, and K. C. Lohmann. 1994. "Evolution of Early Cenozoic Marine Temperatures." *Paleoceanography* 9: 353–387.

INDEX